SpringerBriefs in Applied Sciences and Technology

PoliMI SpringerBriefs

More information about this series at http://www.springer.com/series/11159
http://www.polimi.it

Giuliana Iannaccone · Marco Imperadori
Gabriele Masera

Smart-ECO Buildings Towards 2020/2030

Innovative Technologies for Resource Efficient Buildings

 POLITECNICO
DI MILANO

 Springer

Giuliana Iannaccone
Marco Imperadori
Gabriele Masera
A.B.C. Department
Politecnico di Milano
Milan
Italy

ISSN 2282-2577   ISSN 2282-2585   (electronic)
ISBN 978-3-319-00268-2   ISBN 978-3-319-00269-9   (eBook)
DOI 10.1007/978-3-319-00269-9

Library of Congress Control Number: 2013941934

Springer Cham Heidelberg New York Dordrecht London

Printed on acid-free paper

Springer is part of Springer Science+Business Media (www.springer.com)

Foreword

This book intends to contribute reference knowledge on sustainability in building and construction and is designated to any reader interested in the topic. A prime identified group of readers is students in Building Engineering and Architecture, i.e. in short those that will have the most significant impact on the values, attributes and the performance of our future built environment.

The essential knowledge platform to the book is provided by the European Strategic Support Action project "Sustainable Smart-ECO buildings in the EU," *Smart-ECO*, which was performed in the period 2007–2010. With funding from the EU FP6, DG TREN, 12 partners representing research, industry and the international R&D association, CIB undertook the challenging tasks to develop and anchor a Vision on Sustainable Buildings for the period 2010–2030, to scrutinise research and market to determine what Innovations (technical and nontechnical) that could be useful for realising this Vision, and to identify those elements that have the highest potential impact, all while anchoring the work and all findings with a carefully selected wide-ranging international group of expert stakeholders involved in various aspects of the built sector.

The *Vision* for sustainable Smart-ECO buildings outlines an ambitious direction of development for the time-frame up to 2030. It is based on the state of the art in international standardisation as the situation appeared during the completion of the project, together with findings of other performed R&D activities. With internationally agreed documents such as the CIB Agenda 21 on Sustainable Construction, the ISO General Principles of Sustainability in Building Construction as important background documents, and with evaluation and assessment of the international stakeholders the Vision gained supportive attention of the international standardisation community as well as the UNEP-SUN programme.

The Vision and the resulting requirements were in a defined process condensed to an approach to identify the *Innovations* to support the implementation of the Vision. Also in a foreword, it may be vital to signal the awareness of the project group of the utmost challenge imposed by the task to at a given slot of time identify Innovations that may impact on at least a 20-year period. However, when doing this we did the best possible, also with an awareness that this in fact mirrors the natural effects of the long-lived nature of buildings; what we realise and build today will have due consequences many years to come.

The *Evaluation of the Innovations* intended to identify those innovations understood to have the largest potential for a development in line with the Vision. The *Evaluation* was a multi-criteria approach.

In the Evaluation process, the *Anchoring with stakeholders* was essential. Some 230+ technical experts, industrialists, property developers, material experts, architects, builders, demolition companies and educationalists were engaged in following, guiding and scrutinising the work. This was a tedious but rewarding process. The below figure seeks to describe the project framework and the process specially focussing the involvement of stakeholders.

Dear reader, welcome to engage in the great challenge, *Sustainable Construction*. As the Co-ordinator of Smart-ECO, former President of CIB during the period when CIB established the Agenda 21 on Sustainable Construction, and former chairman of the ISO standardisation activities on Service Life Planning in Building Construction I sincerely hope your views and opinions, ambitions and actions will take this important subject area to new levels.

KTH, Sweden   Christer Sjöström
                                                                Professor Emeritus

Acknowledgments

This book is an evolution of the work carried out by the research group of Politecnico di Milano coordinated by Marco Imperadori in the framework of the European funded project Smart-ECO. We were privileged to be part of the Smart-ECO consortium, a cluster of researchers and professionals who work internationally in the field of the design and construction of sustainable buildings and built environment. In particular our thanks go to: Christer Sjöstrom of KTH (Project Coordinator), Wolfram Trinius of Buro Trinius (Project Manager) and Stefano Saldini of Mace (WP3 leader). Moreover, we would like to thank: Jean-Luc Chevalier and Alexandra Lebert of CSTB, Amber Stevens of PricewaterhouseCoopers, Hywel Davies of CISBE, Gian Carlo Magnoli Bocchi of Mission Carbon Zero, Gurvinder Singh Virk of KTH, Roode Liias of Tallinn University of Technology, Leo Bakker of Tno, Guri Krigsvoll of Sintef, Santiago Gonzalez Herraiz of European Commission, Bill Porteous and Wim Bakens of CIB.

Many thanks to Vijayalaxmi J. of Anna University in Chennai for her participation in developing some of the early materials of this book and her proof reading.

We also are grateful to everyone who supported the book with their images: Blauraum Architekten (Hamburg), Atelier2 (Milan), Aiace (Milan).

We could not have started our interest in sustainable building design and the related innovative technologies without the teachings of Ettore Zambelli, passionate designer and professor at Politecnico di Milano, whose work is also reflected in these pages.

Last but not least, the time and effort we dedicated to the writing of this book are the result of the love and support of our families: Gian Piero and Claudia Imperadori, Pietro and Libera Masera, Ciro and Carla Iannaccone, Daniela, Sergio and our beloved little Brayan, Irene and Maia, Irene.

Contents

Chapter 1
Smart-ECO: A Real Vision for Energy Efficient Architecture Towards 2030

Abstract In order to spread a shared vision of sustainable energy efficient buildings in the horizon 2020–2030, the European Commission funded the research project Smart-ECO—Sustainable Smart-ECO-Buildings in the EU. The challenge was to identify the technological and non-technological conditions to design and implement buildings that balance aspects as diverse as aesthetics, cost-effectiveness, accessibility, functionality, history, production, safety, energy efficiency and reduced impact on the environment. This chapter highlights the general approach of the research work that supported the construction of a Smart-ECO vision. The research's results have proved that future buildings already exist and good examples of design and innovative technologies could be easily replied in full respect of the EU roadmap.

Keywords Nearly zero energy buildings · 20-20-20 target · Sustainable buildings · Climate interactive house · Sustainable building technologies

In order to reach the 20–20–20 target in 2020 the European Commission (European Commission 2007) attributes a strategic importance to energy efficiency measures in the construction sector. That is due to the recognized huge impact of buildings on total energy consumption and carbon emissions, estimated respectively at 40 and 20 % of the total (European Commission 2009).

Design strategies and building technologies aiming to reduce significantly the energy efficiency of buildings are already available: for this reason the recast of Directive 2002/91 on Energy Performance of Buildings—Directive EPBD 2010/31/EU—sets the target of Nearly Zero Energy Buildings (NZEB) for all buildings built after 2020 (2018 for public buildings). Therefore, buildings are required minimising their energy needs and producing most of their energy from on-site renewable sources.

Although several cases show the viability of the NZEB standard—in particular in the residential sector—in most of the Member States these practices still need to be brought to the market and to common practice. Exceptions can be found in those countries—such as Germany, Switzerland and Austria—where strategies for

---

This chapter is written by Marco Imperadori

© The Author(s) 2014
G. Iannaccone et al., *Smart-ECO Buildings Towards 2020/2030*,
PoliMI SpringerBriefs, DOI 10.1007/978-3-319-00269-9_1

energy efficient building (with respect to the energy demand for winter heating) are a standard practice.

Regarding to the need to spread a shared vision of sustainable energy efficient buildings, the European Commission funded the research project Smart-ECO—Sustainable Smart-ECO-Buildings in the EU (FP6-2005-TREN4-038699 period 2008–2010) aiming to set a standard of European sustainable buildings in the horizon 2020–2030. The challenge was to find the conditions to design and implement buildings that balance aspects as diverse as aesthetics, cost-effectiveness, accessibility, functionality, history, production, safety, energy efficiency and reduced impact on the environment.

The research group was composed by 14 partners and was coordinated by prof. Christer Sjöström from BMG Gävle (Sweden). Partners of the research project were BMG Gävle (S), CSTB (F), Tallinn University of Technology (EST), Servitec (I), TNO (NL), Sintef (N), Fachhochschule Südwestfalen (D), Endoenergy (UK), Politecnico di Milano (I), Hywel Davies Consultancy (UK), Mace (UK) and CIB (NL).

The Smart-ECO vision relies on a wide group of stakeholders to build a consensus-based framework. Stakeholders were involved in the research project through questionnaires and specific workshops held during the periodic project meetings. The Smart-ECO vision sets demanding future standards for buildings and highlights a wide range of issues that need to be addressed over a building lifetime.

In particular Smart-ECO project:

- defined a vision of European sustainable buildings in 2020–2030;
- identified the innovations (technologies and process) required to implement the vision;
- evaluated the most promising innovations;
- and disseminated the results among the operators of the construction sector in view of the 2020 energy efficiency targets.

So, what will buildings be like in 2020–2030? A very important part of the research project was the definition of a "vision", shared by the research group and the stakeholders alike.

Although sustainability of buildings has been studied for a long time, there is no universally accepted definition of a "sustainable", "ecological", or whatever definition may be used, building. Only recently, ISO finally defined a standard containing a shared definition of "sustainability" applied to buildings (ISO 15392:2008 Sustainability in building construction—General principles). Along with this very important starting point, the research considered other important aspects defining the state of the art about sustainability, such as CIB's Agenda 21 (CIB 2001), the various national legislations, certification and evaluation tools, etc.

According to the resulting vision, a Smart-ECO building in 20 years should:

1. be designed from a lifecycle point of view;
2. be constructed with limited resources and minimised energy consumption and waste production;
3. have minimised operational complexity while allowing easy monitoring of technical and environmental performances;

4. be adaptable to changes in capacity, type of users and performance requirements;
5. include local issues in all aspects of design, construction, use and dismantling;
6. facilitate ease of dismantling—reuse, recycle, restore.

As anticipated, the recast of the Directive on energy performance of buildings, defines pretty clearly the requirements on energy efficiency and carbon emissions from 2018 on. Other aspects are not yet defined by regulations or best practices, but were deemed significant for the evolution of buildings (and architecture) in the next 20 years. While most of these issues may look common sense or obvious, the real challenge for the European Commission is to have these concepts transferred to the market, making them current practice for decision-makers, designers, clients, construction companies, etc.

The approach, as a method, towards sustainability doesn't exclude any material (is steel less natural than glulam wood?) and therefore also polymeric ones have been taken in consideration (Giulio Natta, professor at Politecnico di Milano, was the only Italian Nobel Prize in Chemistry that awarded in 1963).

The projects introduced as case studies in this chapter, which are suitable to show in advance the Smart-ECO roadmap, have in common the building technology: the Structure/Envelope technique that allows easy assembly and disassembly operations, high flexibility and performances. In general, they all have the same DNA, i.e. they're conceived with a "cyclical design" approach: not only they react differently to the cycle of season and to outer climate, but also they are a stage of a life cycle: "from cradle to a new cradle" (Figs. 1.1 and 1.2).

What characterizes a Smart-ECO Building? First of all, the project includes more complex analysis and specific design issues related to the target performances and the context in which the building is located. We can imagine these buildings as *filters* between the macro environment outside and the micro living environment inside. The three dimensional border has to be designed carefully and take into account many factors like thermal resistance, thermal delay, acoustics, seismic, solar radiation, and so on. All these can be easily managed in those climates that are constant during the year or require a more detailed analysis in those climates that have a significant difference between winter, summer and the middle seasons.

As a result, the future is not a *Passiv House* but even beyond it: an *Active House* or a *Climate Interactive House*. This concept opens the way for strategies to design and build *Cyclical Buildings*, i.e. buildings that are able to act/react in harmony with the nature and the seasons and where every part is potentially re-cyclable. Every component can be inserted in a chain of use which is always designed in order to create less enthropy and more efficiency.

Smart-ECO pointed out the necessity to increase density and concentrate building volumes, both in new or retrofitted buildings, in order to improve energy performances (compared to single houses) and also energy distribution/sharing/self production. A desirable scenario would be to move towards smart-grids enhanced by information and communication technology where very efficient buildings are connected into a network producing more energy than they need and this energy can be shared and managed on demand (Lund et al. 2014).

Fig. 1.1 High energy efficient building in Stezzano, Italy (Architect: Atelier2—Gallotti e Imperadori Associati). The energy demand of the house is 22 kWh/m^{2}y

Following the previous consideration, we can finally compare three experimental buildings—three houses of small dimensions, which already embody Smart-ECO concepts: Kingspan Lighthouse in London, Darmstadt team-Solar Decathlon 2007 in Washington and Casa E3—Vanoncini in Bergamo.

They are the result of integrated design approaches merging academic research and professional activity, and they are early examples of very energy-efficient buildings, some of them already compliant with the 2020 requirements of the EU.

They reveal how is possible to deliver Smart-ECO buildings adopting different strategies, different materials and also different shapes and design approaches. They are three different buildings which express three different architectural approaches, although sharing the same principles and performance targets.

Example: Lighthouse. Designer: Sheppard Robson and ARUP The Lighthouse has been built at BRE's Innovation Park in London with the target Level 6 "net zero-carbon for homes in use", the highest level of the government's Code for Sustainable Homes (CSH), mandatory in UK by 2016. The building was designed by Sheppard Robson and ARUP for Kingspan. It is characterized by a mainly blind south façade, a roof facing south accommodating a PV array and solar hot water heating collector for both radiant heating and solar cooling (Fig. 1.3). The building openings face east and west and can are protected in summer by external movable shutters. The

Fig. 1.2 Residential complex in Torre Boldone, Italy, special award for the environmental sustainability and bio-ecological and green building solutions at 9th IQU Prize—Innovation and urban quality (Architect: Atelier2—Gallotti e Imperadori Associati)

building envelope is constructed using high performance SIPS (structurally insulated panel based system), which provides a high level of thermal insulation and performance, and a ventilated façade of lattice wood (Fig. 1.5). The double-height living space enables the natural ventilation through the roof windows on the top level under the wind catcher (Fig. 1.4). In order to absorb the room heat, internal room surfaces are "thermally heavyweight", made of dense cement fibreboards and plasterboard embedding phase change materials (PCM). The building also includes a wastewater management system for re-use: rainwater for the garden and washing machine, shower and bath water for the WC (Figs. 1.3, 1.4 and 1.5).

Example: Solar Decathlon. Design: TU Darmstadt (team leader: prof. Manfred Hegger) Designed by the team of students from TU Darmstadt, leaded by prof. Manfred Hegger, the house has won the prestigious Solar Decathlon 2007 in Washington and then re-built in Germany. It's a compact parallelepiped volume that can be subdivided in 3 volumes. The structure is prefabricated, made by a timber frame with a multi-layer envelope. Vacuum insulation panels (VIP) panels provide super-insulation and plasterboard with embedded PCM create "thermally heavyweight" internal surfaces. Windows opening are limited to the south (triple glazing gas filled window) and north façade. The building has glazing that look south and

Fig. 1.3 The Lighthouse at BRE's Innovation Park in London

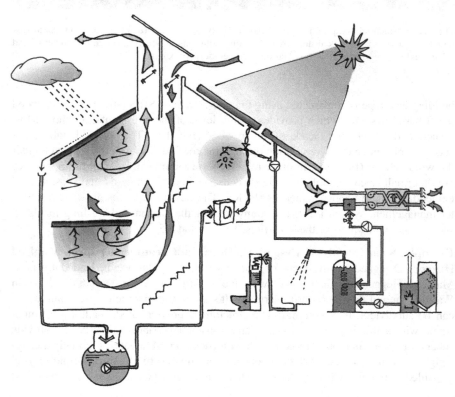

Fig. 1.4 Energy efficiency concepts adopted in the Lighthouse

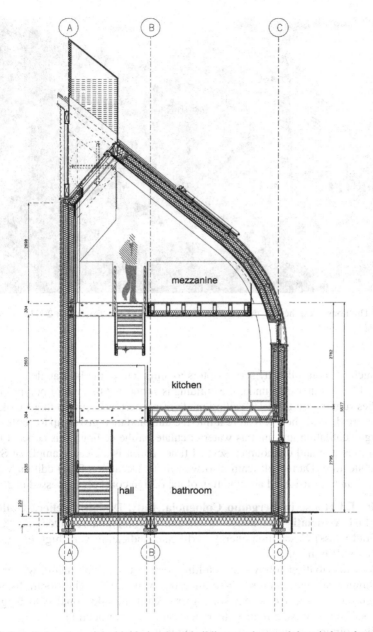

Fig. 1.5 Detailed section of the highly insulated building envelope and the wind catcher

north (quadruple glazing gas filled window) (Fig. 1.6). The timber façade accommodates amorphous photovoltaic modules. On the south façade, a glass covered porch with photovoltaic modules is shaded by louvered panels containing photovoltaic

Fig. 1.6 The house designed by the TU Darmastadt team at Solar Decathlon 2007 competition in Washington (DC)

strips which are controlled by servo motors to aim them at the best angle toward the sun (Fig. 1.7). A water tank under the building is connected to a net of polypropylene microtubes running on the false ceiling made of plasterboard embedding PCM. This help a absorb daytime heat and store it into the tank and then give it up to cooler night time purge ventilation leaving this water circulate on the ceiling. This concept house has been mounted and dismounted several times and it is a clear example of Smart-ECO architecture. Darmstadt team also won Solar Decathlon 2009 edition with an house that can be considered an evolution of the design concepts expressed in 2007.

Example: E3 House in Bergamo Colognola, Italy. Design: Atelier2-Gallotti e Imperadori Associati This house in Bergamo Colognola received the "Classe Oro" (Gold Class) CasaClima energy certificate, indicating very high energy performance (6 kWh/m^{2}a).

The indoor distribution follows the building orientation, maximising winter solar gain, summer shading and north-south cross ventilation. The south façade is largely glazed to improve winter solar gains, with the additional help of glazed sunspaces. Timber louvers reduce the risk of summer overheating.

Innovative solutions adopted in this building include:

- possibility of North-South cross ventilation;
- exploitation of natural light;
- lightweight steel structure;
- 100 % recycled insulation panels (polyester);

Fig. 1.7 A detail of the glass covered porch with photovoltaic modules

- conservatories for passive solar gain;
- phase change material (PCM) embedded in plasterboard panels to store heat in the conservatories;
- solar panels for water heating;
- energy-efficient condensation boiler;
- radiant floor panels for winter heating;
- mechanical ventilation with heat recovery.

The building optimizes the land use (see Sect. 2.2.1) as he substitutes an existing building, an old warehouse for agriculture (the typical double-pitched roof was imposed by the Bergamo Town Hall so as to follow the volume of the existing building). The building envelope has very high insulation properties (U value = 0.09 W/m² K) and a good thermal inertia according the Italian

Fig. 1.8 The south façade of the E3 house in Colognola

regulations. Different insulation materials are used: rock wool, wood wool, polyester wool obtained by recycled plastic bottles, thin multi-foil insulation materials derived from aerospace technologies (see Sect. 3.1.1.1). The building is very close to the Milano Orio Al Serio Airport, and therefore the use of different insulation materials guarantees high acoustic performances.

On the south façade conservatories are placed that collect energy in winter and store it in the back wall made by plasterboard with embedded Phase Change Materials (see Sect. 3.1.1.4). The conservatories have smart openable roof windows allowing for the stack effect (Figs. 1.8 and 1.9).

Due to the high level of airtightness, in winter or in summer the building is hermetic and uses mechanical ventilation with total heat recovery and dehumidification. In the middle seasons, depending from environmental conditions of heat and humidity, natural ventilation or hybrid ventilation guarantee the internal comfort.

In conclusion, Smart-ECO has shown and proved that future buildings are already present and that good examples of design and innovative technologies could be replied and could define a trend to follow in full respect of the laws and strategies of EU. Mankind has clearly affected, and is affecting, the planet and there is a clear need of change in the way to design and build. Everything changes and

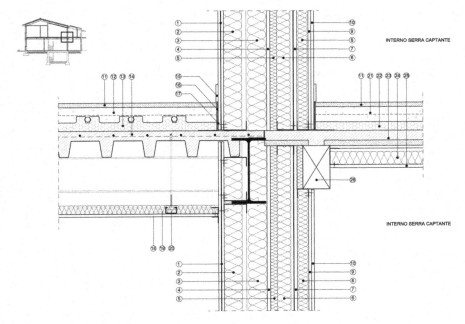

Fig. 1.9 A detail of the wall inside the conservatories made by plasterboard with embedded PCM

the very famous sentence by Charles Darwin "It will not be the strongest or the most intelligent to survive but the one able to change" can drive us towards a better future and a new horizon for construction: from *Stone Age* to *Solar Age*.

References

CIB (2001), Agenda 21 on sustainable construction, CIB Report Publication 237. http://cibworld. xs4all.nl/dl/publications/agenda21.pdf. Accessed 10 Dec 2013

European Commission (2009), EU energy and transport in figures, Office for Official Publications of the European Communities, Luxembourg. http://ec.europa.eu/energy/publications/doc/2009_energy_transport_figures.pdf. Accessed 10 Dec 2013

European Commission, Directorate-General for Energy and Transport (2007), 2020 vision: saving our energy. http://ec.europa.eu/energy/action_plan_energy_efficiency/doc/2007_eeap_en.pdf. Accessed 10 Dec 2013

H. Lund, S. Werner, R. Wiltshire, 4th generation district heating (4GDH) integrating smart thermal grids into future sustainable energy systems. Energy (in press)

Chapter 2
Holistic Design Applying Innovative Technologies

Abstract This section highlights the importance of the adoption of a life cycle approach in the Architecture, Engineering and Construction industry so as to deliver real sustainable buildings. The different impacts of buildings on the environment are considered: depletion of resources, alteration of local microclimate, wastes, storm water management, etc. Their assessment adds specific tasks to the design and decision-making process that can be effectively managed in the framework of an integrated building design. This is a model radically different to the approach that became consolidated in the 20th century design process and, for this reason, requires new operational tools. Building Information Modelling (BIM) and other methods of integrated project delivery may help firms and organizations to enhance efficiency throughout the building process and target sustainability considering the life cycle perspective at all levels.

Keywords Integrated building design · Life cycle · Building information modelling · Holistic design issues · Environmental impact

2.1 Process and Management of Information

2.1.1 Whole Building Design

Buildings are required to satisfy ever more complex requirements than in the past and their life cycle is increasingly entering the design process, as a consequence not only of environmental concerns but also of a growing interest in continuously rising follow-on costs and therefore in the use-phase of buildings (Nemry 2010). The diffusion of innovative models of project financing such as public private partnership (PPP), private finance initiative (PFI) or build-operate-transfer (BOT), based on long term rents and evaluation of performances over several years, has also promoted the further development, systematization and application of building

---

This chapter is written by Gabriele Masera

G. Iannaccone et al., *Smart-ECO Buildings Towards 2020/2030*,
PoliMI SpringerBriefs, DOI 10.1007/978-3-319-00269-9_2

operating costing and life-cycle costing. In fact, interest in whole-life costing has vastly increased due to the recognition that first costs are a very small part of the overall cost of buildings. Nevertheless, the growing interest in sustainable building design can be attributed to the general recognition that there are direct economic benefits from sustainable building: from real savings and by improving the financial performance of a building. Costs in the life cycle of a building depend on what extent it can fulfil today's use requirements and those of the future. That follows the society's expectations and matches the wider concept of sustainability. Furthermore, clients and users are becoming more aware of how the cost of operation, servicing and maintenance of a building plays a significant part in the overall economy, possibly exceeding the capital outlay within a few years of occupancy (Tsai 2011).

The Smart-ECO vision sets demanding future standards for buildings and highlights a wide range of issues that need to be addressed over a building's lifetime. The challenge is to design and implement buildings that balance aspects as diverse as aesthetics, cost-effectiveness, accessibility, functionality, history, production, safety, energy efficiency and reduced impact on the environment (Prowler 2012).

These complex objectives should also be harmonised to address different expectations (Hegger et al. 2008): users and occupants expect unpolluted indoor climate, good lighting and ventilation, view of the outside, flexible internal layouts and low running costs; clients and investors expect dependable schedule of costs for construction and operation of the building, and trouble-free use of their building; public authorities demand that buildings be safe and not to cause unnecessary social costs and negative environmental impacts.

2.1.2 Life Cycle Approach to Design and Construction

Addressing sustainability in buildings leads to an expansion of the spatial and temporal system boundaries. As a matter of fact, a building's long-term impacts on the environment are important if advances in building design are to be sustainable. Designers and stakeholders must consider the impact on conditions "upstream" and "downstream" of the building in what is commonly referred to as a cradle-to-grave approach. This objective introduces specific tasks during the design and decision-making process.

An integrated approach to building design allows thinking holistically about a project, considering the life cycle perspective at all levels and balancing the different interests involved in the development, design, use and management of buildings. This concept is essential to the definition of a Smart-ECO building.

As a matter of fact, different stakeholders handle the different planning activities of a building's life cycle with disjointed, short-term and incomplete interaction among them. Long-term sustainability suffers as a result, leading to increased negative effects on the environment. Every action taken with respect to a building generates impacts within the building, in the surrounding community and beyond. These have real consequences on the well being of people, land, air and water, plants and animals, and generations to come. These in turn have consequences on the future of the building. In this sense, the inter-linkages between each stage of the process (design, construction, use, maintenance, demolition) together with the broader, collective environmental effect of these components have to be considered jointly and in the long-term.

The goal of holistic design is to convert buildings and operations into carbon-neutral architecture where the above issues are addressed while making the energy used in their design and operations a sum net-zero gain (Hernandez and Kenny 2011). It is a cradle-to-grave analysis of all embodied energy in the making of a building, the use and the recycling of that product so it can be used again instead of becoming waste. The goal is to eliminate waste entirely, by ensuring that all constituent parts can be re-used in full, in a closed cycle, with no loss of quality.

Reaching the ambitious goals of the Smart-ECO vision requires an approach to the design process that is radically different to the approach that became consolidated in the 20th century. This typically saw an architect design a building, hand it over to the engineers who may substantially change the design to suit issues of cost, structure or energy thereby losing the opportunity of integration of architecture and engineering. Any attempt to implement an integrated process needs to address the complexity of building. Any building is the result of hundreds of linear processes that are completed in order to obtain the final result. Poor integration results in components being designed in isolation and value being lost. Poor communication results in errors, omissions, and assumptions that result in over-sizing systems, redundancy and gaps in knowledge and performance analysis.

Integrated building design can be defined as an advanced design-and-build model to govern energy, resources and environmental quality decisions. The management of a complex decision-making process requires the efficient exchange of information so that every professional can take informed decisions knowing their impact on the building as a whole.

The aim of integrated planning is to achieve a holistic assessment of individual aspects to gain both horizontal (interdisciplinary) and vertical (life cycle-related) integration. It is therefore possible to introduce new findings and requirements into the planning process from the beginning and implement efficient optimisation techniques such as feedback, simulation of variants. The use of building information modelling (BIM) helps the implementation of sustainable design at the early stage of a project when the total capital costs of a building can be influenced effectively (Jrade and Jalaei 2013).

The holistic consideration of different objectives also means that quantifiable variables (e.g. energy efficiency) become strictly related to other qualitative aspects (e.g. form of the building, image, traditional building techniques). An integrated planning process should attempt to guarantee that both quantitative and qualitative aspects are considered over the entire lifespan. A good integration happens through a continuously dynamic, iterative process. All issues are addressed early and kept in play for as long as possible so that connections and relationships can be optimised. The integrated design process—or integrative design process (7group et al. 2009)—can be described as a repeating pattern of research/analysis and team workshops. Sustainability brings new challenges for engineers dealing with the task of minimizing the use of natural resources while maximizing the use of renewable resources, optimizing, at the same time, whole engineered system's long-term functionalities. In general, while there's a lack of a solid foundation to connect the traditional building engineering disciplines, relevant vast knowledge accumulated in the fields of environmental ergonomics, building materials, and durability analysis is not yet applied consistently during the building process. Different examples show how integrate life-cycle considerations into the building process, from transactional models to systems' integration through the application of the principles of building science (Mora et al. 2011).

The considerations above show that a drastic shift is required in the way projects are managed and calls for better integration of specialists involved in the process and, very basically, for a more organic sharing of information inside the design process among specialists that take part in the project and outside the process in interaction with clients and stakeholders.

This leads to integrated project delivery that engages contractors and manufacturers early in the design process to add practical value to developments.

This process is currently applied mostly to large and complex projects but has application for a much wider range of developments. Integrated project delivery calls for both a different approach to project management and specific tools that make the sharing of information efficient and reliable.

2.1.3 Managing the Process: Operative Tools and Techniques

Design tools are fundamental to the integrated buildings design; they range from simple calculation procedures to complex simulation models to estimate performance. Tools will provide different degrees of confidence depending on the quality and amount of the input data, the complexity of the calculations and the skill of the user. Many advanced simulation tools are available but most are less useful as early design tools as they aim for great degree of accuracy and thus require larger amounts of data and user time. Simple, effective tools can make designs more visual (3D) or show how proposed changes to a design may affect performance.

Good design decision at an early stage can demonstrate a constructive approach to planning requirements and greatly reduce the risk of costly later revisions. Using preliminary tools, the design team can start to make informed decisions on building location, orientation, built form distribution, materials, types of servicing systems and fuels. As design information is often limited at the pre-planning design stage, checklists and good practice guides are commonly used to identify design considerations that will influence the eventual performance of the development.

Considering, for example, the field of environmental assessment, there is a huge number of different tools developed all over the world for both building products and whole building assessment frameworks, covering different phases of a building's life cycle. All these tools can be analysed and categorised, but their comparison is difficult, if not impossible (Haapio and Viitaniemi 2008).

The improved interoperability of design software will have significant benefits for overall project delivery. Developments in information and communication technologies give an opportunity to improve the way energy simulation tools are used to measure the energy performance of buildings throughout their life cycle (Crosbie et al. 2011).

Traditionally, the interdisciplinary collaboration among all the professionals involved in the design and construction of buildings was limited by the exchange of 2D drawings and documents, although many designers are used to manage 3D models and applications for visualization and design development. The widespread development and use of CAD packages and the increased level of automation in construction processes provide encouraging motives for the exchange of 3D data in the multidisciplinary collaboration: Building Information Modelling (BIM) is envisaged to play a significant role in this transformation, even though its adoption as multi-disciplinary collaboration platform still encounters many obstacles (Arayci et al. 2011; Singh et al. 2011).

Building Information Modelling (BIM) and other methods of integrated project delivery may help firms and organizations to enhance efficiency throughout the building process (Barlish and Sullivan 2012; Sacks et al. 2010). BIM is currently the most common denomination for a new way of approaching the design, construction and maintenance of buildings. It has been defined as "a set of interacting policies, processes and technologies generating a methodology to manage the essential building design and project data in digital format throughout the building's life-cycle" (Succar 2009).

A Building Information Model (BIM) is a digital model with a related database in which all the information about a project (design, fabrication, construction and project management logistics) is stored. It can be 3D, 4D (integrating time) or even 5D (including cost). A BIM offers a more effective way of working, reducing transaction costs and opportunity for errors to be made. For example, the UK Government has stated that from 2014 onwards all contracts awarded will require the supply chain members to work collaboratively through the use of "fully collaborative 3D" BIM.

BIM creates a complete digital representation of a building, including physical attributes, geometric form, material descriptions, and thermal and structural behaviour. It grows throughout design, informs construction, and continues to serve facility managers during post-occupancy operations. It has been accepted as key to integrated project delivery (IPD) in which the owner/designer/builder team cooperates in shared risks and rewards (Novitski 2009).

An analysis on 35 case studies, selected among completed construction projects that implemented BIM, allowed to explore the extent to which the use of BIM has resulted in reported benefits. The data obtained from the case studies suggest that BIM is an effective tool in improving certain key aspects of the delivery of construction projects: cost reduction was the one most positively influenced by the implementation of BIM followed by time reduction, improvement of communication, coordination and quality (related to both the quality of the process and improved documentation and enhanced designs). The negative benefits of BIM implementation are focused on software or hardware issues that relate to the need for a general up-skilling within the sector and stakeholder engagement activities to allow key actors to get used to a new way of working (Bryde et al. 2013).

As advanced tool for the management of integrated building design process, BIM adoption in the Architecture Engineering and Construction industry (AEC) would require a change in the existing work practice. A different approach to model development is needed in a collaborative setting where multiple parties contribute to a single shared model (Gu and London 2010). Nevertheless, recent studies reveals that it is possible to align the functionality of BIM based software applications with generally established working methods without the need to implement work process changes at the same time (Hartmann et al. 2012).

2.2 Holistic Design Issues

Aside from operational energy needs, buildings impact on land use, materials, surrounding environment and water. For each of these areas, holistic design solutions and evaluation tools are highlighted along with relevant issues and innovations.

2.2.1 Land Use

Land is a finite resource, nevertheless urban areas are constantly expanding in terms of space and density. The increased proportion of land used by buildings (cities) induces at least three orders of problems:

- Competition with other uses (production of food, production of energy and preservation of biodiversity);
- Increased use of fossil fuels required for individual travel and consequent depletion of resources and carbon emissions that means increased waste and pollution;
- Increased area of impermeable surfaces.

The global population is becoming more concentrated in urban areas. At the moment, 80 % of people in Europe live in towns and cities and spend most of their time indoors. The percentage of urban population worldwide is increasing. At a global scale, 75 % of the world's population is expected to live in urban areas by 2050. Historical trends show that European cities have expanded on average by 78 % since 1950 but the population has grown by only 33 %. A major consequence of this trend is that European cities have become much less compact; the space consumed per person in the cities of Europe during the past 50 years has more than doubled.

During the 10 years period 1990–2000, the growth of urban areas and associated infrastructure throughout Europe consumed more than 8,000 km^2 (a 5.4 % increase during the period), equivalent to the entire territory of Luxembourg (EEA

2006). This is equivalent to the consumption of 0.25 % of the combined area of agriculture, forest and natural land. Although these changes may seem small, it has to be considered that urban sprawl is concentrated in areas where urban growth was already very high during the 1970s and 80s.

The sprawl of cities also creates competition with other land uses. The United Kingdom has a land area of 24.3 Mha. Of this, around 4.6 Mha is arable cropland, 12.4 Mha grasses and rough grazing, 2.9 Mha forest and 4.3 Mha are categorised as urban land and other agricultural land (Dunster et al. 2008). It is estimated that a maximum of 4 Mha could be spared for energy crops, with additional tension between the production of crops for heat and power and for transport. The emphasis that many governments are putting on biofuels has recently shown the potential problems induced by the finiteness of land.

On the other hand, the increasing distance of residential districts from working and leisure areas induces widespread use of cars, use of fossil fuel and related pollution. On average, traffic accounts for nearly 1/3 of energy consumption in Europe, the most frequent journey being that between home and work.

There is a direct correlation between urban density and the average energy consumption per capita, showing that compact cities do have lower energy costs, in particular because most of the journeys can be made on foot, by bike or with public transport. Indeed, some studies suggest that, in a Central European city like Hamburg, a relatively old building located in the city with an energy consumption for space heating of 200 kWh/m^2 year and a floor area of roughly 150 m^2 is roughly equivalent to a Passivhaus (energy consumption lower than 15 kWh/m^2 year) in the suburbs requiring to drive a car to work for 10,000 km/year (Hegger et al. 2008).

These arguments support the idea that regulating authorities should aim at the densification of cities, not in the core but in the less dense peripheral areas, promoting the re-use of existing urban voids and brownfields and promoting refurbishment or substitution of existing buildings (Ardente et al. 2011), stopping, on the other hand, sprawl and the depletion of the countryside.

The optimal density level seems to lie between 50 and 150 persons per hectare while higher densities do not deliver significant reductions in energy consumption because they require high-rise buildings that are energy-intensive (both in operational and embodied energy terms). A good mix of uses and functions and high-quality urban spaces mixing built-up and open areas would deliver the following advantages: shorter routes to work and leisure activities, less traffic and pollution, no duplications of services and infrastructures (typical of functionally segregated cities), higher potential for social integration. This framework has as a consequence that rural areas can re-gain a defined function as suppliers of food and energy. Thus, the action of authorities should be focussed not only on the use of parcels of land, but also on possible ways to increase density, where desirable, expanding existing buildings with extra storeys, horizontal expansions, extensions, perimeter block infills, inner courtyard infills.

These measures, where desirable and compatible with existing infrastructures and open areas, can also be the occasion to pay for the energy rehabilitation of energy-intensive buildings of the 20th century.

Example: Treehouses Bebelallee, Hamburg The complex of residential buildings, built in 1959, had a typical yellow brick masonry façade Fig. 2.1. The client wanted to double the dwelling surface, at the same improving the energy performances of the complex and halving CO_2 emissions. In order to preserve the original character, the architects decided to apply a composite insulation layer on the existing outer envelope, finished with bricks in various tones of brawn and grey Fig. 2.3.

The 2-storey extension (walls and ceilings) has a light timber framed prefabricated structure Fig. 2.4 that was assembled on site in a relatively short time, with minimal noise and disturbance for the inhabitants that could stay in their apartments during construction works. The external cladding is made of untreated wood-shingles that recall the surrounding trees (Fig. 2.2).

The outcome is a reduction by 62 % of energy consumption: the gross annual consumption of primary-energy is 81 kWh/m^2 against the previous 217 kWh/m^2.

Architects: blauraum architects partnership, Hamburg
Client: Robert Vogel GmbH & Co., Hamburg
Year: 2010

2.2.2 Impact on Surrounding Environment

Buildings have multiple effects on the natural systems at different scales due to the large amount of materials they use and to the embodied and operational energy they require. During their life time, buildings also impact on their immediate surroundings due to alterations to the properties of surfaces. In dense urban areas, these effects build up and create significant alterations to local microclimate known as urban heat island (UHI) effect. UHI makes temperatures in cities higher

Fig. 2.1 The Bebelallee complex before renovation

Fig. 2.2 The complex after renovation with the 2-storey building extension

Fig. 2.3 A section of
one building block and a
particular of the external
façade with the two different
cladding systems

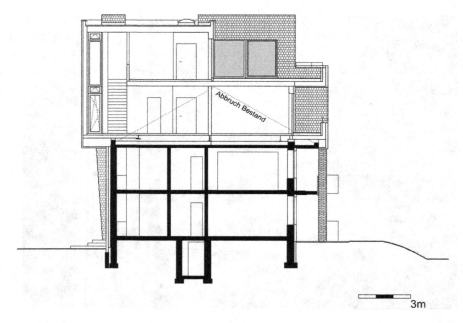

Fig. 2.4 A section of one building block and a particular of the external façade with the two different cladding systems

than in the countryside inducing higher energy consumption for cooling and discomfort for people. The reasons for this are (Kleerekoper et al. 2012):

- The night cooling of surfaces by radiation to the sky is blocked by the presence of buildings; the heat is intercepted by the obstructing surfaces in the street canyons, and absorbed or radiated back to the urban tissue.
- Built areas change the thermal properties of surface materials (absorption, emissivity, thermal capacity).
- The evaporation from urban areas is decreased because of 'waterproofed surfaces'—lack of "natural" soil and vegetation. As a consequence, more energy is put into sensible heat and less into latent heat.
- If tall buildings are present, they provide multiple surfaces for reflection and absorption of sunlight in low albedo materials.
- Air pollution in the urban atmosphere absorbs and reemits long-wave radiation to the urban environment; moreover, combustion processes, such as traffic, space heating and industries, release anthropogenic heat.

Studies suggest that, in metropolitan urban heat islands, air temperatures in summer can be up to 4 °C warmer than the surrounding countryside (Arrau and Peña 2011), due to the extremely large heat capacities of built surfaces that make up for a huge reservoir of energy. The UHI may also affect the wind patterns, development of clouds and fog, humidity, and the rate of precipitation. Other effects include increased energy consumption for cooling (discharging even more waste heat to the outdoor air, although this appears to be a secondary cause of the UHI effect) and influence on the health and welfare of residents (Urban Heat Island 2009).

As Smart-ECO strategies suggest the importance of compact cities, it is important to approach the question holistically and avoid potential negative effects deriving from the densification of urban areas. There are two main strategies that may help reduce UHI effects: (1) cool roofs and paving materials; (2) green roofs (Berndtsson 2010).

Both strategies are relatively common and well experimented, but it is their large-scale application that can make a difference at the urban scale.

The use of light-coloured or cool roofs leads to measured energy savings (cooling) of 10–40 % in buildings with significantly large roofs. For example, the extension of roofs with high reflectivity to all of the US would lead to overall savings of US\$750 m with related lower carbon emissions.

Green roofs have several positive effects in the urban setting in addition to reduce urban heat islands (Carter and Keeler 2008): to improve stormwater management by reducing surface runoff and improving water quality, (Mentens 2006), to reduce building energy consumption by cooling roofs during summer months (Palomo del Barrio 1998), to reduce noise and air pollution (Bradley Rowe 2011) and create habitats for certain plants and animals and thereby improve urban biodiversity. Moreover, analysing energy saving, construction, replacement phase and surface albedo, both white and green roof result in less impact than the black roof (Kosareo and Ries 2007; Susca 2009; Susca et al. 2011).

Studies on the UHI in the city of Milan showed that, during the last century, winter temperatures increased by 2 °C, and in summer the difference between diurnal and nocturnal temperatures decreased. Substituting existing roofs with green roofs in the industrial areas, a potential decrease of air temperatures of 2 °C was estimated (Poli et al. 2007). Smart-ECO innovations should be directed towards the implementation and diffusion of such systems at a large scale in urban areas.

Example: The R.I.E. Index The City of Bolzano in northern Italy introduced the R.I.E. (Riduzione dell'Impatto Edilizio—Reduction of Building Impact) Index that is intended to be a numerical index of the environmental quality. It is applied to a building lot to assess its environmental quality considering soil permeability and green extent.

$$\text{RIE} = \frac{\sum_{i=1}^{n} Sv_i \frac{1}{\Psi} + (Se)}{\sum_{i=1}^{n} Sv_i + \sum_{j=1}^{m} S_{ij} \, \Psi}$$

where:

Sv_i green permeable, impermeable or sealed;
Si_j not green permeable, impermeable or sealed;
$\emptyset$ coefficient of flow;
Se equivalent surface of trees.

It expresses the ratio between the elements modifying the land use and the management of stormwater. The modification of the land use could be positive (improved water collection) or negative (smaller water run-off). The higher is the RIE index, the better is the management of the land with regard to the quantity of infiltrated storm water and to the benefits for the micro-climate and the environment.

RIE Index varies from 0 to 10. RIE = 0 corresponds to a totally sealed surface (for instance, an asphalted parking area); RIE = 10 corresponds to a totally free surface (for instance a forest).

The calculation of RIE Index is compulsory in Bolzano since 2004 for both new and existing buildings and also for all the cases including external surfaces (roofs, terraces, courtyards, etc.). RIE = 1, 5 is needed for production areas and RIE = 4 for residential areas.

A simple RIE calculation sheet is available on the website of the City of Bolzano (www.comune.bolzano.it).

2.2.3 Materials and Waste

The construction sector is the second largest consumer of raw materials after the food industry. Therefore, the choice of building materials has a considerable effect on the environmental impact not only and can promote or prevent subsequent usage. For this reason, in 2003 the European Commission adopted the Communication on Integrated Product Policy (IPP) with the aim of promote the reduction of the environmental impacts from products and services throughout their life cycle. This aspect is increasingly important as buildings use larger quantities of materials to achieve 'zero energy' targets in operation (Hernandez and Kenny 2010). Generally, energy certification schemes don't include aspects related to the life cycle of buildings and it has been demonstrated that, in some cases, this may give rise to the contradiction of obtaining a better energy classification, while producing a higher energy consumption or more CO_2 emissions in global terms (Zabalza Bribian et al. 2009). Therefore a life-cycle perspective should be included within building energy assessment and rating methods (Hernandez and Kenny 2011; Verbeeck and Hens 2010).

Indeed, the more the operational energy needs decrease, the more it's important to consider the energy for material production and its recycling potential (Blengini 2009).

Several tools (Ness et al. 2007; Finnveden and Moberg 2005; Rossi et al. 2012) and databases—such as ECOINVENT or ATHENA—are currently available as filters in the process of materials' selection. Nevertheless, certifications and ecolabels reflecting life cycle assessment can be useful starting points.

A life-cycle analysis looks at material supply chains to reveals the energy-related and ecological effects of materials and creates the basis for selecting building materials responsibly. A life cycle assessment (LCA) is an evaluation of the relative "greenness" of building materials and products.

Considering the embodied energy in materials reveals the importance of life cycle approach. Every use of material should result in a closed material life cycle. The procedure for the subsequent usage is critical for the ecological value, the preservation of raw material and the energy stored in the material. The main options for designers are:

- **Reuse** Designates the subsequent usage of complete durable building products; design for reuse implies that joints between components and materials must enable easy dismantling for further reuse or replacement.
- **Recycle** Designates the recovery of raw materials, in whole or in parts, for the new production of the same materials when their re-use is not possible.

The reuse of building components is site-specific and time-dependent and requires acceptance that the design and construction process may need to change (Gorgolewski 2008).

The most important measure in facilitating future recycling efforts is to use recyclable materials, to avoid materials that contaminate each other, and to avoid construction designs that are difficult to disassemble (Thormark 2002).

Landfill should be considered only if subsequent usage is impossible. In fact, this form of waste treatment involves high costs and high land consumption.

2.2.3.1 Life Cycle Assessment Methodology

The LCA methodology is internationally standardised (ISO 14040—Environmental management—Life cycle assessment—Principles and framework) and recognized as the most effective and the only holistic tool to measure environmental impacts. LCA is important to sustainable design of buildings because it takes into account a range of environmental impact indicators, such as embodied energy and global-warming potential, and facilitates impartial comparisons of materials, assemblies and entire buildings (Sartori and Hestnes 2007). It is true that LCA of entire buildings has mainly been performed by researchers and few professionals in the building sector and its most common building-related application is the comparison of the environmental impacts of different building materials (Cabeza et al. 2014; Bribián et al. 2011), but it is equally true that the demand for LCA is expected to increase. Moreover, its application in the building sector is particularly difficult, due to the intrinsic complexity of the building itself (Li 2006): the variety of products it is made of, the many impacts it generates during its life cycle, the length of its life and the difficulty to forecast its use and maintenance during its service life and disposal or reuse opportunities after more than 50 years.

As pointed out in the Sect. 2.1.2, the different stages of a building's life cycle are handled by different stakeholders with disjointed, short-term and incomplete links to each other. Different models and approaches have been developed and carried out, in the two last decades, by several researchers to cope with this complexity and the literature review shows that there are still different opened critical issues for the identification of a valid methodology (Peuportier 2001; Malmqvist et al. 2011). The main problem that always arises from these experiences is the lack of inventory data or their poor reliability to give clear and doubtful results. In fact, although a large amount of data is available worldwide, their use needs systematic reviews in order to tune up them with the many differences encountered locally (e.g. the incidence of the transportation in the production process or the different mix energy for the different regions) (Blengini and Di Carlo 2010). Therefore an uncritical use of these data might compromise the effective measure of the sustainability and, eventually, its achievement (Pittau et al. 2011).

2.2.3.2 Embodied Energy

"Grey energy" or "embodied energy" is all the energy required to extract, transport, manufacture, assembly and install a building material as well as that required to its disassembly, deconstruction and decomposition.

Analysis on case studies showed that life cycle energy use of buildings depends for the 80–90 % on the operating energy and for the 10–20 % on the embodied energy (EE) of the buildings (Ramesh et al. 2010; Sharma et al. 2011). As already discussed, the significant reduction of the operating energy through the use of passive and active technologies may lead to an increase in embodied energy (due to of the energy consumed for the fabrication of insulation, triple glazing, equipment, etc.). Linking energy and life cycle assessment allows balancing embodied and operation energy when choosing e.g. insulation thickness or window type (Peuportier et al. 2013).

Even if a real and consistent comparison between materials and products is difficult, due to inaccuracy and variability of data in the different databases and methodologies that are used to determine the embodied energy of building materials (Dixit et al. 2010, 2012), in general, wood (from sustainably managed forests) is considered the building material with least embodied energy.

Example: €CO_2—Wood in carbon efficient construction "Wood in carbon efficient construction" was a research project, coordinated by Aalto University, focused on the demonstration of the positive effects on climate of using wood in construction. The findings are the result of a large transnational European research project involving twenty organizations from five countries: Austria, Finland, Germany, Italy and Sweden (Kuittinen et al. 2013). Even if the current normative policy framework in these emerging matters is still under development, the findings of €CO_2 scientifically prove that there are convincing advantages and potentials for using wood in construction to mitigate climate change, whereas the forests are managed so as to maintain or increase forest carbon stocks (Fig. 2.5).

The research filled in some knowledge gaps by applying advanced methods for determining the carbon footprint of wooden buildings during their full life cycles. A carbon footprint analysis of wooden buildings is more complex than that of many other products due to the dynamics of forest growth and the variety of co-products involved. Geographical and country-specific differences have significant effects on the carbon footprint of wood-based products (Villa et al. 2012). Climatic differences have an impact on forest species and management methods, and country-specific energy mixes have an impact on CO_2 emissions (ECO2 2011).

Fig. 2.5   An Italian case-study in the €CO₂ project: Progetto C.A.S.E. L'Aquila (Luigi Fragola and Partners, Studio Legnopiù Srl)

The largest part of embodied energy is usually that in load-bearing structure (56 %). Other building components groups with large amounts of embodied energy are the façade (14 %) and the internal fitting-out (20 %). It has been demonstrated that about 10–15 % of the embodied energy of a conventional building can be recovered through recycling (Thormark 2006). Even if there isn't an internationally accepted method for assessing and comparing the recycling potential of materials, a recent study suggests the potential recycling energy as a factor for assessing the recycling value of materials separate from embodied energy (Saghafi and Teshnizi 2011).

2.2.3.3 Service Life Planning

Several major concepts come into play when using LCA and thinking from a life-cycle perspective as service life and durability. Service life refers to how long the building is being designed to last. It is most important to establish an explicit projected service life for the building. Durability functions within the context of service life. Any building system will last only as long as the entire combination of components of the system last. In theory, durability describes the potential for a building material to maintain the functions assigned to it for a certain length of time. To determine durability, we combine the lengths of time for which building components survive undamaged in a defined usage situation.

Example: LMS-Bygga Villa tool The Life cycle Management System (LMS) Bygga Villa tool estimates service life and maintenance intervals of different building parts and systems based on environmental-dependent degradation models. The tool is not an advanced simulation tool but a user-friendly, simple tool that will give the private client a hint of future need of maintenance and long term performance of the building and its heating system (Fig. 2.6). The LMS-Bygga Villa tool is built on database technique and is divided into two interfaces: an administrative interface and a user interface. The administrative interface is restricted to those who are responsible for maintaining and upgrading the tool. This part of the tool consists of a catalogue of building types, building components and different heating systems, building materials and corresponding damage functions. The administrative part is linked to the user interface in which the user will be able to define a building, its location and calculate the service life for each building component/system. The location of the building is selected from a list of cities or from a Geographic Information System (GIS) map (the latter will be developed in the near future). Each location (city) has a defined environment, based on data from meteorological and environmental institutions as well as from the cities themselves. This data is coupled to the city and stored in the administrative part of the database. The building is composed piece by piece, in a hierarchical order, where each building component and system is selected from the building type catalogue (Hallberg et al. 2008).

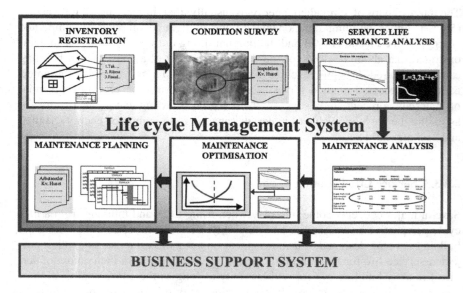

Fig. 2.6 Structure of a life cycle management system and its connection to other business support systems

2.2.3.4 Material Flow Analysis

The Material Input Per Service unit (MIPS) expresses the mass of materials needed to produce a unit of a service, i.e. the materials inputs necessary for producing and using a building product or component. The total material input is the "ecological rucksack", i.e. the mass (in kilograms) of those natural resources that have been removed from their original places in ecosystems and are required for the complete production process of a product, excluding its own weight.

When the MIPS is reduced, the productivity of natural resources is enhanced. This can be done either by decreasing the use of natural resources (MI) or increasing the number of service units (S) provided by the product. This procedure reveals materials flows and identifies the main resources used per unit of mass.

The use of natural resources (MI) can be cut down e.g. by moving to raw materials and energy sources with a small ecological rucksack (MI factor) improving industrial processes so that they use less raw materials and energy, making products smaller and lighter (preferably without reducing the number of service units) using recycled materials in manufacturing the products or diminishing the need of the transport of raw materials or finished goods.

The number of service units provided by the product can be increased by promoting the longevity of products (by making them durable, timeless, maintainable, repairable, upgradeable, supplementable, reliable, easy to use and reusable) and their multistage use (by making them reusable, multi-purpose, easily disassembled, and simple in terms of their material consistence and structure).

2.2.3.5 Ecolabel—Environmental Labelling

Eco-labels started with products and are evolving into methods of assessing entire buildings. Ecolabelling provides consumers with information about the environmental performance of products. The criteria for awarding an ecolabel are derived following a study of the environmental impacts associated with the whole life cycle of a particular product. In particular, when building products are considered, raw materials extraction, building design, construction practice, building use, and demolition will often have more significant environmental impacts than the manufacturing process itself.

2.2.3.6 Design for Deconstruction and Disassembly

The concept of Design for Deconstruction or Design for Disassembly (DfD) considers future demolition and disassembly of building elements at the design stage of new buildings promoting waste and resource-use reduction. The appropriate use of building technologies and their successful integration into the design process will facilitate an increased reuse of the building components. The difference between deconstruction and disassembly was debated at the "Deconstruction—closing the loop" Conference held in 1999 at the Building Research Establishment (BRE): disassembly is a process of taking apart components without damaging them, but not necessarily to reuse them, while deconstruction is a process similar to disassembly but with thought towards reusing the components (Hurley et al. 2001). One of the main barriers to deconstruction is that buildings are not designed to enable it. Making construction easier—by means of a proper design and selection of building materials and technologies—can contribute to make deconstruction easier. Guidelines for deconstruction (Table 2.1) are already available and primarily concerns with issues of handling, access, and prefabrication (Crowther 2000).

Table 2.1 Principles of design for disassembly (DfD) as applied to buildings (Crowther 2000)

1	Use recycled and recyclable materials
2	Minimize the number of types of materials
3	Avoid toxic and hazardous materials
4	Avoid composite materials and make inseparable products from the same material
5	Avoid secondary finishes to materials
6	Provide standard and permanent identification of material types
7	Minimize the number of different types of components
8	Use a minimum number of wearing parts
9	Use mechanical rather than chemical connections
10	Use an open building system with more freely interchangeable parts
11	Use modular design
12	Use assembly technologies compatible with standard building practice
13	Separate the structure from the cladding, the internal walls and the services
14	Provide access to all parts of the building and all components
15	Use components sized to suit the intended means of handling at all stages
16	Provide a means of handling components during disassembly
17	Provide adequate tolerance to allow for movement during disassembly
18	Minimize types of connectors
19	Design joints and connectors to withstand repeated assembly and disassembly
20	Use a hierarchy of disassembly related to expected life span of the components
21	Provide permanent identification for each component
22	Standardise the parts while allowing for an infinite variety of the building as a whole
23	Use a standard structural grid
24	Use a minimum number of different types of components
25	Use prefabricated sub-assemblies
26	Use lightweight materials and components
27	Identify point of disassembly permanently
28	Provide spare parts and storage for them
29	Retain information on the building manufacture and its assembly process

2.2.4 Water

Water consumption is strictly connected to the level of urbanization in a country. As the land becomes predominately urbanized, it will be more difficult for cities to meet the rising demand for freshwater; in many areas the consumption of water exceeds the ability of the supplying aquifer to replenish itself. At the same time water is a critical energy medium in the climate system.

In the Smart-ECO vision, good building design can greatly reduce the amount of water used and discharged into the sewage with the benefit of reducing sewage system failures caused from excess water overwhelming the system, reduced water contamination caused by polluted runoff and reduced need to construct additional water and wastewater treatment facilities. Efficient water use can also reduce the amount of energy needed to the treatment systems (Cheng 2002), resulting in less energy demand.

2.2.4.1 Total Water Consumption and Role of Buildings

Globally, between 12.5 and 14 billion cubic meters of water per year are available for human use. In 1989 about 9,000 cubic meters was available per person per year; by the year 2025 global per capita availability of freshwater is projected to drop to 5,100 cubic meters per person as another 2 billion people join the world's population. In developing countries the drinking water consumption per person per year is about 450–1,800 cubic meters, in Europe 6,200–9,100 cubic meters.

Buildings have many water demands that vary according to location and building type. Supplying and treating cold water requires a significant amount of energy. For instance, American public water supply and treatment facilities consume about 56 billion kWh per year.

Key innovations should focus on using less water throughout the building lifecycle. In addition new technologies should maximize the water recycling (storm water, grey water, black water) and incorporate green roofs into the project.

Water supply and drainage systems for buildings provide several opportunities for including sustainable solutions, without compromising their overall performances (Jack and Swaffield 2009). Moreover, water-saving devices and solutions can be effectively adopted in existing building, mainly in those countries characterized by water shortage (Liu and Ping 2012).

Options include:

- Incorporate efficiency in construction specifications;
- Adopting alternative design measures performance enhancement and for water and pipework economies;
- Use ultra water-efficient plumbing fixtures and integrate other water-saving devices into buildings;
- Design landscape for water efficiency through use of native plants tolerant of local soil and rainfall conditions that not require permanent irrigation;
- Meter water usage;

- Eliminate leaks;
- Specify water-efficient labelled products;
- Recover non-sewage, roof water, groundwater and grey water for on-site activities use.

2.2.4.2 Storm Water and Wastewater Reuse

In urban areas storm water is generated by rain run-off from roof, roads, driveways, footpaths and other impervious or hard surfaces. Rainwater can be used for toilet flushing, laundries or for watering the garden.

Wastewater (grey water + black water) re-use decreases effluent volumes, reducing the stress on existing centralized wastewater disposal systems, which will work better and last longer.

At an urban scale, waste water treatment plants require an area of 0.5–2 m^2 per inhabitant. A drop in water consumption in conjunction with the smaller household sizes plus the increase in the living space requirements per person have resulted in reduced waste-water flows. The cost of waste-water treatment in large-scale plants is much greater than the cost of treating drinking water. The energy requirements of waste-water plants (due to aeration, circulation and transportation of the waste-water) are also very high. Thus, decentralised waster-water treatment offers a cost-effective alternative.

Technologies for controlling water consumption are only one aspect of maximizing water effectiveness; others are wise management of water sources and even the potential to produce usable water, through treatment technologies or desalination. It is naturally prudent to consider the whole impact of a process such as desalination due to the intensive energy demands.

Innovative solutions isolate usable partial flows from the water by means of materials-flow analysis. It is possible, for example, to use rain water in the building services, e.g. for cooling, with both low-tech solutions for open bodies of water and even high-tech solution for air conditioning systems already available.

Example: Bank of America Tower, New York (USA) The Bank of America Tower is the second-highest skyscraper in New York and the first office tower in the US to be nominated for LEED Platinum, the highest ecological rating of the American Green Building Council (Fig. 2.7). The concept of sustainability also includes minimizing water consumption partly through the use of rainwater.

The majority of wastewater generated in the building is recycled by collecting sink wastewater, coil condensate from the air-conditioning systems, and roof storm water into a series of tanks.

Planting to the flat roof of the podium serves to retain rainwater and prevent excessive heat gains. Rain that falls on the surfaces at the top of the tower is fed into four grey-water tanks at various levels of the building. In the main tank in the basement, water from various sources is collected and used for cooling purposes and for flushing toilets. The installation of waterless urinals brought the greatest overall saving.

Fig. 2.7 The Bank of America Tower, a symbol of sustainable construction in the skyline of New York

2.2.4.3 Embodied Water

The embodied water is a concept analogous to that of embodied energy and embodied carbon. It refers to the cumulative quantity of water needed to produce a product through the supply chain. The embodied water contains both direct and indirect water paths that have not previously being included when considering the water consumption of the construction industry (Crowford 2011). In fact, policies have focused on the operational water use of the built environment, neglecting the embodied water of various goods and services required for construction.

Direct water is that consumed in the main production of the specific product being analysed while indirect water is that used to create and deliver materials and resources that go into the main product. It is harder to define because of the many sources of consumption that may be involved. Concept and methods of embodied energy analysis, derived from life cycle assessment (LCA) and more directly from life cycle energy assessment, can be used to undertake an embodied water analysis (McCormack et al. 2007). Built on the ISO 14000 family of Life Cycle standards, the developing ISO 14046 Water Footprint standard, specifies the principles, requirements and guidelines of assessing and reporting water footprints of products, processes and organizations. ISO 14046 will provide requirements and guidance for calculating and reporting a water footprint as a standalone assessment or as part of a wider environmental assessment.

References

7group, B.G. Reed, R. Fedrizzi, *The integrative design guide to green building* (Wiley, Hoboken, 2009)

Y. Arayci, P. Coates, L. Koskela et al., Technology adoption in the BIM implementation for lean architectural practice. Autom. Constr. **20**, 189–195 (2011)

F. Ardente, M. Beccali, M. Cellura, M. Mistretta, Energy and environmental benefits in public buildings as a result of retrofit actions. Renew. Sustain. Energy Rev. **15**, 460–470 (2011)

C.P. Arrau, M.A. Peña (2011) Urban Heat Islands (UHIs). http://www.urbanheatislands.com. Accessed 27 Dec 2013

K. Barlish, K. Sullivan, How to measure the benefits of BIM—a case study approach. Autom. Constr. **24**, 149–159 (2012)

J.C. Berndtsson, Green roof performance towards management of runoff water quantity and quality: a review. Ecol. Eng. **36**, 351–360 (2010)

G.A. Blengini, Life cycle of buildings, demolition and recycling potential: a case study in Turin, Italy. Build. Environ. **44**, 319–330 (2009)

G.A. Blengini, T. Di Carlo, The changing role of life cycle phases, subsystems and materials in the LCA of low energy buildings. Energy Build. **42**, 869–880 (2010)

D. Bryde, M. Broquetas, J.M. Volm, The project benefits of building information modelling (BIM). Int. J. Project Manage. **31**, 971–980 (2013)

L. Cabeza, L. Rincón, V. Vilariño et al., Life cycle assessment (LCA) and life cycle energy analysis (LCEA) of buildings and the building sector: a review. Renew. Sustain. Energy Rev. **29**, 394–416 (2014)

T. Carter, A. Keeler, Life-cycle cost–benefit analysis of extensive vegetated roof systems. J. Environ. Manage. **87**, 350–363 (2008)

C. Cheng, Study of the inter-relationship between water use and energy conservation for a building. Energy Build. **34**, 261–266 (2002)

T. Crosbie, N. Dawood, S. Dawood, Improving the energy performance of the built environment: the potential of virtual collaborative life cycle tools. Autom. Constr. **20**, 205–216 (2011)

R. Crowford (2011) Life cycle water analysis of an Australian residential building and its occupants. in *Proceedings of 7th Australian Conference on Life Cycle Assessment*, Australian Life Cycle Assessment Society, Melbourne, 9–10 Mar 2011

P. Crowther (2000) Developing guidelines for design for deconstruction. in *Proceedings of the "Deconstruction—Closing the Loop" conference*, BRE, Watford, UK, 18 May 2000

M.K. Dixit, J.L. Fernández-Solís, S. Lavy, C.H. Culp, Identification of parameters for embodied energy measurement: a literature review. Energy Build. **42**, 1238–1247 (2010)

M.K. Dixit, J.L. Fernández-Solís, S. Lavy, C.H. Culp, Need for an embodied energy measurement protocol for buildings: a review paper. Renew. Sustain. Energy Rev. **16**, 3730–3743 (2012)

B. Dunster, C. Simmons, B. Gilbert, *The ZEDbook* (Taylor & Francis, Abingdon, 2008)

ECO$_2$ (2011) http://www.eCO2wood.com

EEA (2006) Urban sprawl in Europe. The ignored challenge. http://www.eea.europa.eu/publications/eea_report_2006_10. Accessed 27 Dec 2013

G. Finnveden, A. Moberg, Environmental systems analysis tools e an overview. J. Clean. Prod. **13**, 1165–1173 (2005)

M. Gorgolewski, Designing with reused building components: some challenges. Build. Res. Inf. **36**(2), 175–188 (2008)

N. Gu, K. London, Understanding and facilitating BIM adoption in the AEC industry. Autom. Constr. **19**, 988–999 (2010)

D. Hallberg, J. Akander, B. Stojanović, M. Kedbäck (2008) Life cycle management system—a planning tool supporting long-term based design and maintenance planning. in *Proceedings of 11DBMC international conference on durability of building materials and components*, Istanbul, 11–14 May 2008

T. Hartmann, H. van Meerveld, N. Vossebeld, A. Adriaanse, Aligning building information model tools and construction management methods. Autom. Constr. **22**, 605–613 (2012)

A. Haapio, P. Viitaniemi, A critical review of building environmental assessment tools. Environ. Impact Assess. Rev. **28**, 469–482 (2008)

M. Hegger, M. Fuchs, T. Stark, M. Zeumer, *Energy manual.* (Birkhäuser—Edition Detail, Basel, 2008)

P. Hernandez, P. Kenny, From net energy to zero energy buildings: defining life cycle zero energy buildings (LC-ZEB). Energy Build. **42**, 815–821 (2010)

P. Hernandez, P. Kenny, Development of a methodology for life cycle building energy ratings. Energy Policy **39**, 3779–3788 (2011)

J.W. Hurley, C. McGrath, S.L. Fletcher, H.M. Bowes, *Deconstruction and reuse of construction materials* (BRE, Watford, 2001)

L.B. Jack, J.A. Swaffield, Embedding sustainability in the design of water supply and drainage systems for buildings. Renew. Energy **34**, 2061–2066 (2009)

A. Jrade, F. Jalaei, Integrating building information modelling with sustainability to design building projects at the conceptual stage. Build. Simul. **6**, 429–444 (2013)

L. Kleerekoper, M. van Escha, T. Baldiri Salcedo, How to make a city climate-proof, addressing the urban heat island effect. Resour. Conserv. Recycl. **64**, 30–38 (2012)

L. Kosareo, R. Ries, Comparative environmental life cycle assessment of green roofs. Build. Environ. **42**, 2606–2613 (2007)

M. Kuittinen, A. Ludvig, G. Weiss eds. (2013) Wood in carbon efficient construction—Tools, methods and applications. Hämeen Kirjapaino Oy, Tampere. http://www.eCO2wood.com/2

Z. Li, A new life cycle impact assessment approach for buildings. Build. Environ. **41**, 1414–1422 (2006)

B. Liu, Y. Ping, Water saving retrofitting and its comprehensive evaluation of existing residential buildings. Energy Procedia **14**, 1780–1785 (2012)

T. Malmqvist, M. Glaumann, S. Scarpellini et al., Life cycle assessment in buildings: The ENSLIC simplified method and guidelines. Energy **36**, 1900–1907 (2011)

M. McCormack, G.J. Treloar, L. Palmowski, R. Crawford, Modelling direct and indirect water requirements of construction. Build. Res. Inf **35**(2), 156–162 (2007)

J. Mentens, D. Raes, M. Hermy, Green roofs as a tool for solving the rainwater runoff problem in the urbanized 21st century? Landscape Urban Plann. **77**, 217–226 (2006)

R. Mora, G. Bitsuamlak, M. Horvat, Integrated life-cycle design of building enclosures. Build. Environ. **46**, 1469–1479 (2011)

F. Nemrya, A. Uihleina, C. Makishi Colodel, Options to reduce the environmental impacts of residential buildings in the European Union—Potential and costs. Energy Build. **42**, 976–984 (2010)

B. Ness, E. Urbel-Piirsalua, S. Anderbergd et al., Categorising tools for sustainability assessment. Ecol. Econ. **60**, 498–508 (2007)

B.J. Novitski (2009) BIM promotes sustainability. Practitioners are finding paths to green through interoperable software. http://continuingeducation.construction.com/article.php?L=5&C=516. Accessed 4 Dec 2013

E. Palomo Del Barrio, Analysis of the green roofs cooling potential in buildings. Energy and Buildings **27**(2), 179–193 (1998)

B.L.P. Peuportier, Life cycle assessment applied to the comparative evaluation of single family houses in the French context. Energy Build. **33**, 443–450 (2001)

B. Peuportier, S. Thiers, A. Guiavarch, Eco-design of buildings using thermal simulation and life cycle assessment. J. Clean. Prod. **39**, 73–78 (2013)

F. Pittau, E. De Angelis, G. Masera et al. LCA Based Comparative Evaluation of Building Envelope Systems. in *Proceeding of CISBAT11 International Conference*, Lausanne 2011

T. Poli, L.P. Gattoni, R. Arlunno et al., The influence of albedo of surfaces on microclimatic modifications. New scenarios for Milano. in *PLEA2007—24th International Conference on Passive and Low Energy Architecture Proceedings*. National University of Singapore, Singapore, 22–24, Nov 2007

D. Prowler, Whole building design (2012), http://www.wbdg.org/wbdg_approach.php. Accessed 27 Dec 2013

T. Ramesha, R. Prakasha, K.K. Shukla, Life cycle energy analysis of buildings: an overview. Energy Build. **42**, 1592–1600 (2010)

B. Rossi, A.F. Marique, M. Glaumann, S. Reiter, Life-cycle assessment of residential buildings in three different European locations, basic tool. Build. Environ. **51**, 395–401 (2012)

D.B. Rowe, Green roofs as a means of pollution abatement. Environ. Pollut. **159**, 2100–2110 (2011)

R. Sacks, I. Kaner, C.M. Eastman, Y.S. Jeong, The rosewood experiment—building information modeling and interoperability for architectural precast facades. Autom. Constr. **19**, 419–432 (2010)

M.D. Saghafi, Z.S.H. Teshnizi, Recycling value of building materials in building assessment systems. Energy Build. **43**, 3181–3188 (2011)

I. Sartori, A.G. Hestnes, Energy use in the life cycle of conventional and low-energy buildings: a review article. Energy Build. **39**, 249–257 (2007)

A. Sharma, A. Saxena, M. Sethi et al., Life cycle assessment of buildings: a review. Renew. Sustain. Energy Rev. **15**, 871–875 (2011)

V. Singh, N. Gu, X. Wang, A theoretical framework of a BIM-based multi-disciplinary collaboration platform. Automation in Construction **20**, 134–144 (2011)

B. Succar, Building information modelling framework: a research and delivery foundation for industry stakeholders. Autom. Constr. **18**, 357–375 (2009)

T. Susca, Enhancement of life cycle assessment (LCA) methodology to include the effect of surface albedo on climate change: comparing black and white roofs. Environ. Pollut. **163**, 48–54 (2009)

T. Susca, S.R. Gaffin, G.R. Dell'Osso, Positive effects of vegetation: urban heat island and green roofs. Environ. Pollut. **159**, 2119–2126 (2011)

C. Thormark, A low energy building in a life cycle—its embodied energy, energy need for operation and recycling potential. Build. Environ. **37**, 429–435 (2002)

C. Thormark, The effect of material choice on the total energy need and recycling potential of a building. Build. Environ. **41**, 1019–1026 (2006)

W.H. Tsai, S.J. Lin, J.Y. Liu, W.R. Lin, K.C. Lee, Incorporating life cycle assessments into building project decision-making: an energy consumption and CO_2 emission perspective. Energy **36**, 3022–3029 (2011)

G. Verbeeck, H. Hens, Life cycle inventory of buildings: A calculation method. Build. Environ. **45**, 1037–1041 (2010)

N. Villa, F. Pittau, E. De Angelis et al., Wood products for the Italian construction industry—an LCA-based sustainability evaluation. in *Proceeding of WCTE 2012—World Conference on Timber Engineering*, Auckland, 15-19, July 2012

Water Footprint Network (2008). http://www.waterfootprint.org. Accessed 27 Dec 2013

I.Z. Bribián, A. Aranda Usón, S. Scarpellini, Life cycle assessment in buildings: state-of-the-art and simplified LCA methodology as a complement for building certification. Build. Environ. **44**, 2510–2520 (2009)

I.Z. Bribián, A. Valero Capilla, A. Aranda Usón, Life cycle assessment of building materials: Comparative analysis of energy and environmental impacts and evaluation of the eco-efficiency improvement potential. Build. Environ. **46**, 1133–1140 (2011)

Chapter 3
Innovative Technological Solutions

Abstract This chapter presents the results of investigations about innovations enabling the building sector to meet the requirements originating from the sustainability concept. These innovations, referred to new construction, refurbishment, or both, are typically already experimented or under development, and can be expected to hit to the market in the next few years. Innovations range from energy saving building technologies and energy efficient generation systems. Taken in isolation, the single innovations can be more or less relevant to the implementation of Smart-ECO buildings: it is quite evident that effective results can be obtained only applying sets of coordinated innovations, as there is no one-fit-all solution. Case studies are considered, where appropriate, to illustrate the feasibility of applying innovations in the built environment.

Keywords Energy saving · Thermal insulation · Cladding materials · Glazing · Thermal storage in building fabric · Natural ventilation · Control systems · Natural light · Shading · Renewable energy sources · Building-integrated energy production

3.1 Energy Saving

This section considers innovative technologies in the field of energy saving for buildings with the view that using less energy is a fundamental step towards a smaller environmental impact.

The direction of this report has been shaped by stakeholder input in the Smart-ECO project, including direct survey of industry experts and findings from interactive workshops that took place during the development of the research.

On the topic of energy saving, the key opinions expressed by stakeholders were the following:

This chapter is written by Giuliana Iannaccone

© The Author(s) 2014
G. Iannaccone et al., *Smart-ECO Buildings Towards 2020/2030*,
PoliMI SpringerBriefs, DOI 10.1007/978-3-319-00269-9_3

- the potential of existing technologies is not being fully utilised. Improving existing technologies or developing new ones is important, but the focus needs to be on implementation and integration of existing technologies;
- passive design measures is the area of innovation with the highest impact potential;
- insulation, passive solar design and passive cooling are the best energy saving technologies, followed by lighting and innovative materials;
- buildings should be automated using existing technologies to reduce consumption, integrate systems and inform users so they adapt their behaviour.

Using the stakeholder views as a starting point, this section aims to identify areas of innovation with the greatest potential to impact Smart-ECO buildings in the period 2010–2030.

Understanding the problems

Globally, the operation of building services such as space and water heating, space cooling, ventilation and lighting consumes more than one third of the total energy, with a comparable contribution to greenhouse gas emissions. These figures are set to rise in the future because of the increasing urbanisation of population (Hegger et al. 2008).

While the impact of buildings on the environment is significant, the construction sector also presents the potential for a drastic reduction of energy consumption and carbon emissions through the application of existing strategies and technologies (International Energy Agency 2011). For this reason, the European Commission considers buildings one of the most strategic sectors to achieve the 2020 targets of energy efficiency (Energy Saving Trust 2010; European Commission 2010a; European Union 2011; Eurostat 2013; Directive 2012/27/EU).

One of the biggest challenges on the road to the improvement of energy efficiency of European buildings is the wide variability of climatic conditions around the continent. European climates are determined not only by latitude or altitude but also by proximity to the ocean or inland seas. With such climatic variation within Europe, there are no single design guidelines that could fit all situations, and it is crucial that buildings are climate responsive and site contextual, beyond the simple variation of the insulation levels.

Apart from designing higher quality, site specific buildings to prevent energy loss, the challenge to energy saving is changing consumer behaviour and habit. Using less is about first understanding how much we use and then knowing what changes in use make a significant difference. The huge task of making the building sector carbon-free as a part of the 2050 de-carbonisation plan (European Commission 2011) can be tackled, given the existing and expected technologies for energy generation, only if energy use is drastically reduced. This is clearly stated in the Directive 2010/31/EU on the energy performance of buildings, where the nearly zero-energy building (NZEB) is defined as a "building that has a very high energy performance", while "the nearly zero or very low amount of energy required should be covered to a very significant extent by energy from renewable sources, including energy from renewable sources produced on-site or nearby" (Directive 2010/31/EU). This definition clearly prioritises the efficiency of the system (fabric and services) over the energy production alone (Voss and Musall 2011).

Working towards solutions

Reducing the energy requirements in buildings (mainly those for heating, cooling and lighting) emphasises the role of the building envelope (opaque and transparent) as a filter between the controlled conditions indoors and the ambient outside. The improvement of the building fabric's performances is at the basis of any climate-conscious design approach, independent of the specific context.

Solutions that allow a low or nearly zero-energy building in a cold climate (or, for temperate climates, in winter) are quite consolidated and have already demonstrated some economic and technical viability, including the Passivhaus buildings in Germany and the Minergie-P buildings in Switzerland. Innovations in the next future are likely to be concentrated on the improvement of existing products, and in particular of insulating materials. In heating-driven climates, the basic principles of very low energy (or passive) houses are very good fabric insulation, air-tightness, exploitation of solar radiation and internal gains and controlled ventilation with heat recovery.

Solutions relating to buildings located in warm to hot climates (aiming at summer comfort with low or no purchased energy use) are instead less consolidated and, up to now, have generally been defined on a case-by-case basis. Assessing the behaviour of a building in the warm season, when windows may be open and heat flows may change direction during the day, requires more accurate and time-consuming analyses.

Generally speaking, an efficient building in summer should be able to reduce overheating with an efficient fabric (insulation, thermal mass and shading), eliminate excess heat with natural ventilation and resort to mechanical cooling only when ambient conditions do not guarantee passive comfort inside the building. Achieving this requires close coordination by designers of technology and mechanical systems to avoid under or over sizing the cooling system.

Climate-conscious, energy-efficient design requires the adoption of a number of coordinated strategies, with conflicting requirements potentially arising between heating and cooling seasons. Innovation lies not just in products and components themselves, but also in the way they are combined and coordinated.

3.1.1 Heating and Cooling

3.1.1.1 Thermal Insulation

To reduce the environmental impact on the built environment, it is essential to consider further development of the efficiency of thermal insulation materials based on the fact that:

- insulation has a great potential for reducing CO_2 emissions;
- energy conserved through insulation use outweighs the energy used in its manufacture. Only when a building achieves a low energy standard does the energy embodied in the insulating materials become significant (Sartori and Hestnes 2007);
- the durability of insulation affects its performance e.g. settlement, physical degradation, vapour permeability and air movement.

Note that insulation only provides reduction of heat loss (or gain) through the building fabric. Equally important is the energy lost through ventilation and glazing (hence the need for integrated design).

Since the operational energy of buildings is gradually reducing, as a result of the new energy efficiency standards, the energy embodied in the production of materials and the construction process becomes more relevant. This is why significant effort is dedicated to reducing the life cycle impact of insulating materials and it is reasonable to expect a larger diffusion of products with high recycled content or deriving from natural sources which can be sustainably regenerated.

A further field of innovation is the improvement of materials' thermal performance (i.e., reducing the heat flow per thickness unit). Improving the behaviour of the insulating material (or system) with respect to conduction, convection and radiation would reduce the thickness required to reach high performances. This would benefit refurbishment/retrofitting operations in particular, where the thickening of existing walls would reduce internal floor area or expand the external walls beyond the building boundaries.

Below are some examples of innovative insulating materials or systems aiming to improve performance and minimise space requirements.

Insulating panels with IR shield

Traditional insulation materials resist the heat flow by reducing transmission from conduction and convection. It is possible to influence radiation transmission by reducing the passage of infrared radiation (emissivity). This is obtained applying selective/reflective layers to a material—usually invisible, thin metallic coatings applied by vapour deposition. An example of this can be found in low-emissivity glazing.

As this technology is not dependent on the backing material, it can also be used in the production of insulating materials, e.g. in the form of modified polystyrene. Thermal conductivity values can be reduced by 20 % ($\lambda = 0.040$ W/m K for standard EPS, $\lambda = 0.032$ W/m K for modified EPS).

Another possibility is adding small capsules of graphite in the mix. These are also able to reflect infrared radiation and thus reduce the overall heat flow through the insulating panel.

Aerogel insulation

Aerogel is a manufactured material with the lowest bulk density of any known porous solid. It is derived from a gel in which the liquid component of the gel has been replaced with a gas. The result is an extremely low-density material. Silica aerogel can be micro-encapsulated and integrated in a blanket of reinforcing fibres. The result is a thin (5–10 mm) insulating panel with extremely low thermal conductivity ($\lambda = 0.014$ W/m K at 25 °C average temperature).

The panel can be used as external and internal insulation in walls, roofs, framing and floors.

Vacuum insulation panels

VIP panels have thermal conductivity values that are around 1/10th of traditional insulation materials ($\lambda = 0.003$ W/m^2 K). VIPs are made by sealing the thermal insulation (in general a porous solid with low density and nano-scale porosity) in a barrier film under vacuum.

The very high performance delivered by VIPs in thin layer makes them very interesting for refurbishment operations. Other fields of application include facade insulation where volume limitations or alignments do not allow for the installation of thick external layers.

Disadvantages include cost and durability, depending on the quality and vapour-tightness of the metallic film around the panel. VIPs are currently supposed to last for at least 20 years.

Multi-foil reflective insulation systems
Thin multi-foil insulation is composed of reflective layers—generally aluminium foil—spaced by separating layers like wadding, foam, etc. Its insulating properties are based on limiting reflection (infrared waves), convection (still air is present in the cavities) and conduction. These insulating systems work properly when installed in cavities, alone or as a complement to traditional insulation.

On-site measurements show good insulating performances of these thin compound systems, making them suitable for refurbishment.

3.1.1.2   Cladding Materials

Recent developments in materials and manufacturing processes are delivering solutions that have the potential to transform the building from a net burden on the environment to an active producer of energy and even an environmental filter.

Advancements in nanotechnology and surface science are opening up new possibilities to improve (or find new) performances of traditional construction materials. The past decade has seen a remarkable revolution in our understanding and mimicking of natural processes at a chemical, physical and atomic structural level (Turney 2009). This is translating to innovative products that make buildings more interactive with the surrounding environment and easier to maintain.

Progress in self-cleaning surfaces is significant and is based on three alternative strategies (ibid.):

- easy physical removal of dirt through superhydrophilic films of water which thoroughly wet the surface;
- prevention of residual dirt adhesion with superhydrophobic surfaces where water droplets readily roll off a surface, collecting dust particles on the way;
- removal of organic or biological surface films by photocatalytic oxidation.

The first two strategies are already applied in commercial products such as glass and coatings (e.g. external render and textiles) that can be cleaned under the simple effect of rainwater. The photocatalytic effects of TiO_2 coatings have been applied into a wide range of paints, glazes and cements. Besides being self-cleaning (TiO_2 has superhydrophilic properties), nanoparticulate titanium dioxide is a potent catalyst under near-UV radiation and can thus be used for the control of airborne pollutants (Diamanti et al. 2008). In the future, the use of TiO_2 coatings may find feasible applications for the control of odour in indoor environments provided there is sufficient UV light intensity.

Recent laboratory developments in thermocromic VO_2 coatings could form the basis for thermally responsive materials. Vanadium dioxide undergoes a marked change in optical transmittance and reflectivity in the IR region, associated with an insulator-to-metal phase transition: in principle, such coatings would absorb IR radiation until they reach the transition temperature, and then become strongly reflecting.

Buildings can also contribute to the definition of the microclimate surrounding the construction itself. Until now, most buildings have contributed to the urban heat island effect (UHI), raising the ambient temperatures as a consequence of heat absorption on external surfaces and its release to the outdoor air. Two examples of how buildings can minimise their negative contribution are cool roofs and green roofs.

Cool roofs and paving materials deliver high solar reflectance (the ability to reflect the visible, infrared and ultraviolet wavelengths of the sun, reducing heat transfer to the building) and high thermal emission (the ability to release a large percentage of absorbed, or non-reflected, solar energy) which reduces the heat build-up on external surfaces contributing to UHI (Akbari and Konopacki 2005).

Green roofs are vegetated roof covers with growing media and plants taking the place of bare membrane, gravel ballast, shingles or tiles. A vegetated surface remains at a lower temperature and may include a water retention system to add to the benefit by evaporation from the wet soil (Barrio 1998; Fiori et al. 2013).

These examples show how buildings can become elements that improve the surrounding environment, contributing to the reduction of air pollution, the mitigation of ambient temperatures and the production of energy.

TiO_2 materials and coatings

Titanium dioxide is the most widely used white pigment because of its brightness and very high refractive index. Titanium dioxide is a photocatalyst under ultraviolet light. Recently it has been found that titanium dioxide, when spiked with nitrogen ions or doped with metal oxide like tungsten trioxide, is also a photocatalyst under visible and UV light.

Titanium dioxide is added to paints, cements, windows, tiles, or other products for its sterilizing, deodorizing and antifouling properties.

Cool roofs

The outer layer surface temperature remains close to the air temperature compared to a standard, dark cladding. The benefits lie in a reduced heat flow in the building with consequent savings on cooling and a reduction of the urban heat island effect. With lower temperatures both indoors and outdoors, user comfort also increases.

In the temperate climates that are typical of most of Europe, the cooling benefits of a highly reflective roof surface far outweigh the potential winter month heating benefits of a less reflective, or black, roof surface.

Cool roofs are categorised as:

- *inherently cool roofs*: these roofs are covered with light-coloured paint or plastic membranes (vinyl) that achieve very high reflectance values;
- *coated roofs*: technologies are available that make standard cladding products more reflective. Products include shingles, tiles, tinted paints, membranes and metal roofing;
- *green roofs*: these are a specific type of cool roof and they are treated separately.

Cool roof solutions are currently rated in the US under programs such as Energy Star, Cool Roof Rating Council, Green Globes and LEED.

The benefits of a cool roof system are more significant in buildings that have large roof to floor area ratios. This makes them less appealing to high-rise buildings, although there are still beneficial effects on user comfort.

Green roofs

Green roofs increase thermal and acoustical performances of the roof, help manage storm-water reduction, filter dust particles in the air and reduce stress on the waterproof membrane. Less quantifiable effects include psychological well-being of users and providing a habitat for wildlife species.

Green roof vary in number of layers and layer placement, but are characterised by a single to multi-ply waterproofing layer, drainage, growing media and the plants, covering the entire roof deck surface.

There are two main types of green roofs:

- *Extensive*: They have thinner and less number of layers so they are lighter, less expensive and require very low maintenance. Extensive green roofs are built when the primary desire is for an ecological roof cover with limited human access. The minimum growing media or soil substrate varies from 5 to 15 cm. Fully saturated weight ranges from 45 to 100 kg/m^2. Low growing, horizontally spreading root covers the ground with plant heights generally 40–60 cm.
- *Intensive*: They look like traditional roof gardens because a much wider variety of plant material can be included, since growing media depths are increased. The growing media generally ranges from 20 to 100 cm, depending on the loading capacity of the roof and the architectural and plant features desired. Fully saturated weights start from 350 kg/m^2. Lightweight solutions that do not significantly increase the structural load on existing buildings make the application of green roofs possible also in refurbishment operations (Fig. 3.1).

Fig. 3.1 The roof of one of the buildings at the Politecnico di Milano campus was retrofitted with eight different types of green roofs, which are now being monitored for physical and thermal performances (R. Paolini, GreenLab, Dipartimento ABC, Politecnico di Milano)

3.1.1.3 Glazing

Low-energy, low-carbon construction concepts call for improved efficiency of the glazed parts of the envelope. In most of Europe, these should allow for passive solar gain during the heating season, while reducing heat losses and avoiding (or reducing) overheating in summer (Gonzalo and Haberman 2006).

Heat transfer by conduction and convection should be reduced by means of improved U-values for glazing and frames, while the control of solar radiation entering the building can be obtained either through shading or the energy transmission properties of the glazed system, which may include materials with variable transparency.

Below are some innovations concerning the transparent components of the building envelope.

Vacuum glazing
Ultra-efficient glass has been available for some time now as triple glazing, and commercial applications of quadruple glazing can prove feasible. The insulation effect at the transition from double to triple (or quadruple) glazing is the outcome of greater system thicknesses, with higher weight and smaller light transmittance. In the case of vacuum glazing, a significant improvement over traditional insulating glass is achieved by obviating the thermal conduction caused by the gas in the cavity between the panes.

Using this form of construction, even with double glazing, excellent insulation values of $U_g = 0.5$ W/m^2 K can be achieved with system thicknesses of less than 10 mm. In view of the relatively low weight, appropriate forms of frame construction are necessary: slender, lightweight and with high insulation properties.

The production of vacuum insulation glazing depends on evacuating the gas from the cavity between the panes. As a result of the external atmospheric pressure that is caused by the evacuation of the cavity, the panes of glass are subject to a load of 10 t/m^2. To prevent the panes being pressed together, small supports or spacers must be incorporated at regular intervals in the cavity.

Glass with translucent insulation
To reduce heat exchange between different sides of a glazing system, it is possible to fill the cavity with translucent insulation. This reduces convection losses due to air movements in the cavity, while still allowing diffuse (but not direct) radiation to enter the building.

One possible solution is the use of acrylic capillary elements, with tubes containing still air installed perpendicularly to the glass. The geometry of tubes allows more energy to enter the building in winter (low sun) and to reduce overheating in summer (high sun).

Another possibility is the use of translucent aerogel. Aerogels are good thermal insulators because they almost nullify the three methods of heat transfer (convection, conduction and radiation). They are good conductive insulators because they are composed almost entirely from a gas, and gasses are very poor heat conductors. Silica aerogel is especially good because silica is also a poor conductor of heat. Carbon aerogel is a good radiative insulator because carbon absorbs the infrared radiation that transfers heat at standard temperatures.

While allowing diffuse radiation into the building, translucent insulating materials do not allow views outside.

Glass including PCM
The use of phase change materials (PCM) in a glazing system can confer to a transparent element a significant heat storage property. When hit by the sun, PCMs absorb energy and

(at the melting temperature) change their state. In the process, they store latent heat while remaining at a constant temperature. During winter nights, this energy can be released to the rooms of the building if an appropriate thermal resistance is provided on the outer side of the glass (e.g. a cavity with a glass pane). Summer radiation can be excluded by means of shading systems or prismatic glass included in the glazing system.

As PCMs are translucent, their presence in the cavity still allows diffuse light to enter the building.

In principle, translucent elements containing PCMs can also be adjustable, in order to regulate the behaviour of the envelope according to season or time of the day.

Glass with variable solar transmittance
Performances required from glass may change drastically according to time of the day or season. Active thin films on glazing are seen as a very attractive approach to solar control and daylight regulation in highly glazed buildings. While external shading devices are usually the best approach to solar control, there are many reasons why their application may be considered undesirable such as capital cost, aesthetics, maintenance, structural restriction and wind loading, especially for high-rise buildings. The use of highly reflective glazing would be an alternative, but this can create glare issues for the exterior surrounding.

A potential solution would be the use of glass with active layers that can modify the light and energy transmission properties of the façade. Examples of this technology include glass that change its properties when a voltage is applied (electrochromic glazing, suspended particle devices), and glass that change its properties in response to an environmental signal (thermochromic glazing, electrochromic glazing). Some companies expect to provide systems with adjustable opacity (not just clear /reflective) in the near future.

Critical aspects of smart glass include installation costs, the use of electricity, durability, as well as functional features such as the speed of control, possibilities for dimming, and the degree of transparency of the glass.

3.1.1.4 Thermal Storage in the Building Fabric

The thermal capacity of a building (the amount of heat that can be stored in its elements) defines its dynamic thermal behaviour, that is, how quickly it heats up or cools down when boundary conditions change.

While in traditional buildings, mass is used to reduce the inbound heat flow through thermal lag and decrement, the levels of thermal insulation required by European regulations shifts the importance on the surfaces that exchange heat with the indoor environment. Thermal storage surfaces can absorb heat due to solar radiation and internal gains in summer, thus reducing overheating of indoor air, or absorb heat in winter, releasing it during the night or when the heating system is switched off. Thermal mass is then able to stabilise indoor temperatures, reducing peak loads on systems (Di Perna et al. 2011).

These surfaces are also instrumental in defining user comfort, as operative temperature is defined by both air and surface temperatures. Thus a warm surface will increase comfort in winter, while a cool surface will be beneficial in summer provided heat can be removed during the night with ventilation.

The relationship among climate, insulation levels, thermal mass, dimensions of glazed parts, and so on is specific to each building and the correct assessment of comfort levels requires dynamic energy simulation tools case by case. Buildings with thermal mass can save cooling energy in summer and often appear to be more robust to misuse by the user (Dunster et al. 2008).

Thermal storage can be realised either with traditional materials or with innovative solutions such as phase change materials (PCM). It is also possible to artificially activate thermal masses through the inclusion of radiant systems in slabs and walls.

Below are some innovations related to heat storage in buildings.

TABS—activation of thermal mass

Thermally activated building slabs (TABS) exploit thermal capacity of construction materials to create a heat sink in the floor slab. The use of a large exposed surface improves thermal comfort (increase or decrease of mean radiant temperature).

In summer, the exposed mass of the floor slab can absorb excess heat from the rooms, thus levelling out peak cooling loads. This heat can be removed via night flushing of the rooms (natural or mechanical ventilation with cool outdoor air), or by circulating chilled water in embedded pipes. For summer comfort, the exposed surface should be the ceiling (higher yield). In winter, the exposed slab can soak up free heat from the sun and release it at a later time. The slab can be kept warm with water circulating in the pipes.

While traditional radiant systems work in the superficial layers of building components (floors, ceilings or walls), TABS are based on the activation of thick layers of heavyweight materials. This is possible embedding plastic pipes in the structural concrete slab, that shall remain exposed to the rooms to optimise heat exchange. Other solutions use water pipes integrated in lost form shuttering to activate the concrete mass above. As the thermal response of the system is slow, a management system that is able to predict the required slab temperatures is required. This temperature is defined by internal gains, weather and humidity.

TABS have a specific field of application in office buildings with discontinuous use. This allows for night flushing of the building masses and/or off-peak charge and discharge of stored heat. TABS are generally applied in new construction, as they need integration in the structural components.

Building-integrated PCMs

A phase change material (PCM) is a substance with a high heat of fusion, where melting and solidifying at a certain temperature is capable of storing and releasing large amounts of energy. Heat is absorbed or released when the material changes from solid to liquid and vice versa; thus, PCMs are classified as latent heat storage (LHS) units.

Initially, the solid-liquid PCMs behave like sensible heat storage (SHS) materials; their temperature rises as they absorb heat. Unlike conventional SHS, however, when PCMs reach the temperature at which they change phase (their melting temperature) they absorb large amounts of heat at an almost constant temperature. The PCM continues to absorb heat without a significant raise in temperature until all the material is transformed to the liquid phase. When the ambient temperature around a liquid material falls, the PCM solidifies, releasing its stored latent heat.

Within the human comfort range of 20–30 °C, some PCMs are very effective. They store 5–14 times more heat per unit volume than conventional storage materials such as water, masonry, or rock. Thus, they may be very effective as a heat storage element in buildings in substitution to heavier materials, such as concrete, which only absorb sensible heat.

PCMs can be organic substances, such as paraffin or fatty acids, inorganic (salt hydrates) or a mix (eutectics). While PCMs have been used for a number of years in healthcare, spacecraft and heating/cooling systems, their use as a building-integrated component is relatively new and promising. It is possible to use either a layer of macro-encapsulated PCMs as a building component, or add micro-encapsulated PCM powder to lightweight elements such as boards, blocks, glass etc. The advantage over traditional heat-storage media is that weight and volume are reduced (for comparable performance) and the phase change temperature can be programmed according to specific needs.

Very interesting fields of application include highly-insulated, lightweight buildings with small need for thermal storage, or refurbishment of existing, poorly performing buildings lacking in thermal storage (in this case, mass cannot be added to the existing structure). PCMs can be used to store solar energy during the day, to shave the peak cooling loads in summer, to store off peak energy during the night, etc.

Some studies showed the potential to integrate PCMs in building components that can stabilise indoor temperatures either passively or in synergy with (typically radiant) heating/cooling systems (Imperadori et al. 2006) (Figs. 3.2, 3.3).

The correct use of thermal mass in highly-insulated buildings requires accurate studies, especially in warm climates; thus it is necessary to spread the use of dynamic simulation tools (assessment of dynamic behaviour of building) to ensure correct design of insulation, thermal storage capacity, glazed components and ventilation. A mixed approach could be adopted, whereas, under certain conditions, standard buildings would be audited with steady-state calculations while more complex buildings would be designed using dynamic simulation tools. California pioneered this double prescriptive/performance approach in its 2005 energy regulations, updated with the 2008 standards (California Energy Commission 2008).

Example: Arup Office, Solihull, UK The workplace pavilions are designed to include generous ceiling heights and roof pods for natural ventilation and maximised daylight, occupant control, excellent air-tightness, exposed thermal mass for passive cooling, visual and direct links between all floors and with the surrounding landscape and internal spatial flexibility.

The bespoke timber facade consists of cladding with louver timber shutters in Western Red Cedar controlling solar gain. Members of staff have manual and motorised control of the shutters and windows.

Fig. 3.2 The "PCM blanket" developed in the European-funded project Changeable Thermal Inertia Dry Envelopes (C-TIDE)

Fig. 3.3 The integration of
the "PCM blanket" into a
drywall solution

Some of the innovative solutions include:

- natural ventilation of deep-plan design offices;
- low energy design avoiding the use of CFCs and HCFCs;
- use of recycled and non-processed materials;
- implementation of an environmental management system;
- a green travel plan;
- use of life cycle and environmental analyses to influence the design (Figs. 3.4, 3.5).

3.1.1.5 Natural Ventilation for Comfort and Overheating Control

Natural ventilation has three quite different functions with relative benefits: to supply fresh air for health, to control the build-up of heat in the mild or hot seasons, and to improve the feeling of comfort for users of the building. As energy conservation issues imply that new buildings are airtight and refurbished ones have reduced leakage rates, air change is still needed for the well-being of users. Thus, the required amount of fresh air should be supplied by controlled ventilation

Fig. 3.4 Arup Office: An external view of the building with the wind cowls on the roof for natural ventilation (Peter Cook)

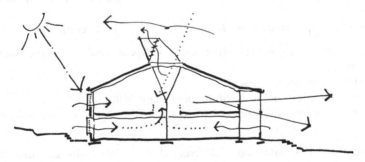

Fig. 3.5 Arup Office: A scheme of the passive energy strategies (*right*) (Arup Associates). *Design* Arup Associates; 2001 (first phase), 2007 (second phase)

systems that in winter are also able to recover heat from exhaust air. While mechanical systems are mature, the challenge is how to deliver the required air change rate while reducing energy consumption from fans and pumps.

When ambient temperatures are mild, natural ventilation can be exploited to dissipate excess heat from the building and to improve comfort. The feeling of comfort improves with open windows and air movement; under natural ventilation conditions, users can still feel comfortable at higher temperatures than in sealed buildings (Brager and de Dear 1998a).

In some seasons, natural ventilation can thus reduce the use of HVAC systems during the day (although it may not completely replace it in all climates). Moreover, it can be beneficial during the night, when it can be used to cool down the components of the building if temperature variations between day and night are large enough.

Building design strategies and components that foster natural ventilation should be implemented in new and refurbished buildings. Due to rising temperatures around Europe (Rubel and Kottek 2010), the latter are under particular stress, especially in those countries where buildings were not designed for hot summers (e.g. traditional homes with large glazing and small vents). Often, it is possible to exploit natural phenomena such as wind forces and stack effect to activate air flows in buildings through carefully placed vents, passive wind cowls and solar chimneys. At a larger scale, an atrium can also be used to temper climate. Atria are intermediate spaces, typically glazed, that rely exclusively on passive behaviour to provide sufficient comfort conditions. Atria exploit direct solar gain in winter and allow for abundant ventilation and shading in summer.

Natural ventilation alone may not always be sufficient to deliver the expected air change rates and indoor air quality (especially in large buildings with high occupancy rates), or it may not even be beneficial if ambient temperatures are too high. This may happen, in particular, in temperate climates that experience extreme hot or cold conditions. It may then be necessary to develop hybrid approaches, allowing for the use of natural ventilation when ambient conditions allow, and of mechanical heating/cooling in peak periods of the year, when temperatures outdoors are not comfortable or when natural drivers of ventilation are too weak.

There are multiple motivations for the interest in hybrid ventilation:

- the likelihood of a positive response by occupants and a positive impact on productivity;
- reduced environmental impact;
- increased robustness and/or flexibility and adaptability;
- commercial factors (the prospect of lower investment costs and/or operation costs).

Hybrid ventilation is based on a different design philosophy and expectations about its performance cannot be the same as for mechanical ventilation. Energy performance targets and comfort requirements must be different. Cost comparisons between hybrid and mechanical systems should be done on a life-cycle cost basis rather than simply on an initial capital cost basis, because of the different design approach, and hence the different balance between initial, running, maintenance and disposal costs (Abrams 1986).

Buildings with hybrid ventilation often include other sustainable technologies e.g. daylighting, passive and natural cooling and passive solar heating. For these buildings, energy optimization requires an integrated approach in the design of the building and its passive and mechanical systems, allowing for the smooth switch between passive and active ventilation, or even the combination of the two (e.g. active supply, passive extract).

The following are examples of innovations that promote natural ventilation in buildings. The correct design of natural ventilation and its interaction with architecture and mechanical systems requires a thoroughly integrated approach and detailed analyses of the natural phenomena taking place in the building (dynamic energy simulation tools, CFD if required, etc.). While these devices have already been applied to buildings in use, guidelines and practical indications are needed to promote their use and overcome lack of knowledge or experience.

Solar chimneys

A solar chimney is a way of improving the natural ventilation of buildings by using convection of air heated by passive solar energy. In its simplest form, the solar chimney consists of a chimney painted black. During the day, solar energy heats the chimney and the air within it, creating an updraft of air in the chimney. The suction created at the chimney's base can be used to ventilate and cool the building below, if ambient temperatures allow. There are a number of solar chimney variations, but basic design elements are:

- a solar collector area—this can be located in the top part of the chimney or can include the entire shaft. The orientation, type of glazing, insulation and thermal properties of this element are crucial for harnessing, retaining and utilizing solar gains;
- main ventilation shaft—the location, height, cross section and the thermal properties of this structure are also very important;
- inlet and outlet air apertures—the sizes, location as well as aerodynamic aspects of these elements are also significant.

The design of a solar chimney needs a thorough coordination with the rest of the building (design of air flows, inlets, thermal mass, user control) and accurate energy calculations, as natural ventilation may even be undesirable on hot days (i.e., when solar radiation and thus the suction effect are higher).

The benefits of a solar chimney are:

- improved ventilation rates on still, hot days;
- reduced reliance on wind and wind driven ventilation;
- improved control of air flow though a building;
- greater choice of air intake (i.e. leeward side of building);
- improved air quality and reduced noise levels in urban areas;
- increased night time ventilation rates;
- ventilation of narrow, small spaces with minimal exposure to external elements.

Passive wind cowls

A wind cowl is a passive heat recovery ventilation system that supplies and extracts air to and from a building. The cowl works like an active ventilation system in that it has dedicated inlet and outlet ducts and a heat recovery system but instead of using electrical fans to drive the air flow it uses the wind to create both positive pressure at the inlet and negative pressure at the outlet ensuring a throughput of air for no electrical input.

In low wind conditions it will continue to produce reasonable ventilation levels through stack effect. The use and economic viability of this kind of systems depends on the wind available at the specific location.

Use of atria for climate tempering

Atria are intermediate spaces are typically glazed and not heated or cooled and thus have a range of temperatures between those outside and those inside the climate-controlled envelope.

With a passive behaviour that exploits direct solar gain in winter and allows for abundant ventilation and shading in summer, these atria can effectively reduce thermal stress on the proper envelope of the building and supply warm or cool air for natural ventilation.

Depending on the climate, these spaces may be closed or open. Moreover, atria can play a social role as common spaces that can be used in inclement weather.

Example: new nursery school, Milan The City of Milan commissioned a design for a new childhood centre that would include advanced low-energy and climate-conscious concepts. The design was developed in the framework of a research funded by the Italian Ministry for University and Research about design strategies and technologies for energy efficiency in mild climates.

The design revolved around the idea that it was possible "to heat the whole school with a system sized for a single flat". In the warm season, the building envelope and especially the transparent elements had to be able to modulate the energy and mass flow to avoid overheating (solar gain, shading, ventilation, etc.) and to provide comfort in free running mode whenever possible (Ferrari et al. 2006).

Some of the innovative solutions include:

- thin-film photovoltaic system;
- natural ventilation by cross ventilation and stack effect (a solar chimney at the top of the atrium roof);
- winter thermal comfort conditions are provided by a primary air ventilation system coupled with floor radiant panels. This combination minimises electricity consumption in pumps and fans;
- a groundwater heat pump system supplies both hot water in winter for space heating and DHW and chilled water for cooling during the warmest parts of summer;
- life cycle assessment (Figs. 3.6, 3.7).

3.1.1.6 Control Systems

Mechanical systems have been designed in the past to supply cheaply available energy to buildings with a very inefficient envelope; today instead, design is moving towards a closer integration of high performance building fabric and very efficient mechanical systems (typically of smaller size compared to traditional buildings). This approach exploits the passive performances of the building, leaving to systems the role of refining indoor comfort conditions (thermal comfort, air quality and lighting) through an efficient use of energy. Since the amount of energy required to operate buildings will decrease significantly between 2010 and 2030 and energy prices are very likely to rise (International Energy Agency cit.), it will be more and more important to manage buildings efficiently (e.g., avoiding waste and over-use of systems). One identified challenge, however, is

Fig. 3.6 Nursery school in Milan: The roof with solar chimneys for natural ventilation (AIACE)

Fig. 3.7 Nursery school in Milan: A section of the building, showing how the chimney draws warm air from the *top* of the atrium (*right*) (AIACE).
Design (Arup Associates). AIACE—Ettore Zambelli with Politecnico di Milano; 2011

that buildings with highly insulated envelopes are often more sensitive to misuse by users when overheating can be a concern. The correct use of ventilation and shading systems are then crucial if buildings are to behave as expected.

Building automation is becoming a fundamental element to deliver buildings that live up to expectations and provide indoor comfort with an optimised use of energy. A Building Management System (BMS) is a computer-based control system installed in buildings that controls and monitors the building's mechanical and electrical equipment, and it is most common in large buildings. As a core function in most BMS systems, it controls heating and cooling, manages the systems that distribute this air throughout the building and locally controls the mixture of heating and cooling to achieve the desired room temperature. A secondary function sometimes is to monitor the level of human-generated CO_2, mixing in outside air with waste air to increase the amount of oxygen while also minimising heat/cooling losses. A BMS can also have the crucial role of controlling solar shading systems in order to optimise heating and cooling loads.

Smart BMSs allow for significant energy savings both in new and existing buildings and avoid over-use of energy (Vraa Nielsen et al. 2011). Although until now BMSs were common in large commercial buildings, smaller systems suitable for households are available under the broad definition of "home automation" systems.

Building automation systems have the advantage of making the idea of the "smart building" practicable where a building can react to constantly changing conditions and situations caused by environment and occupants. A BMS takes over the regulating and controlling functions related both to provision of energy inside the building and to the manipulation of the envelope. While this has the potential to deliver considerable efficiency, it is necessary to underline that extensive automation involves some risks, such as vulnerability to technical system or component failures, higher building costs, our growing dependency on technical systems and also on manufacture and maintenance firms (Herzog 2008).

Moreover, research about adaptive comfort shows that users feel more comfortable, and are tolerant of larger temperature variations, if they can define their own environment, for example opening a window (Brager and de Dear cit.; Brager and de Dear 1998b) or defining their luminous environment through the control of shading devices (Galasiu and Veitch 2006). This is already translating in buildings where the envelope, far from being sealed, can be opened to activate natural ventilation when this is effective. Such hybrid buildings, that can work both with mechanical systems and passive natural ventilation, should become more common but pose additional challenges to BMS designers because of the more complex heat exchange mechanisms and the potential misuse by the occupants.

Smart-ECO buildings will require the right balance between automation and grid integration with the flexibility and resilience provided by overall sound building design (good insulation, effective shading and thermal storage capacity, as described in the relevant paragraphs of this book). This brings us back again to the fundamental role of the designers and of a well designed and planned building.

3.1.2 Natural Lighting and Shading

Good daylight levels are essential for comfort of users and for this reason regulations and standards in Europe mandate either minimum daylight levels, or dimensions of windows relating to size of rooms, or maximum distances of workplaces to windows. Good levels of natural lighting inside the building not only save electricity for artificial lighting but reduces the related internal gains.

The issue of natural lighting needs to be gauged against the ingress of solar energy that may be beneficial in terms of winter heating, but may also lead to glare (especially in workplaces) and to summer overheating, especially in buildings with high internal gains such as offices and schools.

As summer temperatures rise throughout Europe (Rubel and Kottek cit.), the control of solar radiation becomes crucial for homes; this is not only in Southern countries where traditional architecture has always been facing this problem. It is thus necessary to spread the use of effective shading systems that control heat gain and glare (adjustable systems allow for precise control in different conditions) while letting natural light in for visual comfort. Solutions range from glazing-integrated shading to complex, adjustable systems that rely on indirect reflections to spread light in rooms. This sort of shading can be very useful also in refurbishment operations, when lighting levels need to be improved. Shading systems may also be an effective surface for energy generation (e.g., PV panels or vacuum tubes), since they typically receive a high amount of solar radiation.

Natural light is even more desirable in difficult conditions such as deep-plan buildings and urban canyons. Solutions that allow transportation of light deep into the building can range from bespoke architectural features such as light shelves to less visible ones such as light pipes (tubes with highly reflective coating), glass fibres, or heliostats (horizontal or vertical). These are also suitable for improving natural lighting levels in existing buildings.

Examples of efficient shading, light control and light transportation are shown below.

Light shelves
A light shelf is an architectural element that allows daylight to penetrate deep into a building, while providing a degree of shading according to latitude and orientation. It is possible to obtain a light shelf placing a horizontal light-reflecting overhang above eye-level. This element should have a high-reflectance upper surface, reflecting daylight onto the ceiling and deeper into a space. Light shelves are generally made of an extruded aluminium chassis system and aluminium composite panel surfaces, but any reflecting surface can work correctly.

Light shelves make it possible for daylight to penetrate the space up to 2.5 times the distance between the floor and the top of the window. Today, advanced light shelf technology makes it possible to increase the distance up to 4 times.

Light shelves can also provide shade near the windows, due to the overhang of the shelf, and help reduce window glare. Exterior shelves are generally more effective shading devices than interior shelves, as they block radiation before it enters the building. A combination of exterior and interior shelves will work best in providing an even illumination gradient. For maximum benefit, perimeter lighting should be controlled by photosensors, with lighting zones to the particular installation.

Louvers and venetian blinds with special shapes

Innovative shading systems use louvers with a complex geometry, allowing for effective control of solar radiation, but redirecting natural light deep into the indoor spaces. This is achieved either through a particular shape of the louver, or incising the surface of the louver itself with an asymmetrical pattern. In both cases, the louvers, whose shape is always location-specific, are designed to stop most of the direct solar radiation when the sun is high (summer), while still reflecting some indirect light. With the low winter sun, more radiation is allowed to enter the building.

These systems may be fixed or integrated in adjustable blinds (venetian blinds) allowing for a degree of regulation from the users.

Glazing-integrated shading

While external shading devices are the most effective for the control of solar gains (heat is stopped before entering the building), it may be advisable to protect them from wind, rain etc. for maintenance reasons.

Louver systems may be placed in glass cavities, where they remain protected from dust, water and shocks, while preserving the innovative concepts about shading and light reflection that were discussed above.

Vertical angle-selective façade for solar control

The idea behind an angle-selective façade (Frontini and Kuhn 2009) is to exploit the thickness and optical properties of glass (different refraction index between glass and air) to control the amount of solar radiation entering the building and the glare conditions, while still allowing views outside and indirect natural lighting. This is obtained by printing, or applying, thin opaque strips on two faces of a laminated glass pane. The system does not include any movable parts and requires the same maintenance as a standard window or glazed façade. Thin, opaque strips are printed, or applied, between the two panes of a laminated glass and on the inner surface of the façade, their height limited (6 mm max) to avoid them distracting from the views out. The gap between the strips should be designed on a case-by-case basis, according to the sun path of the location and the amount of solar energy allowed to enter the building. Additional solar and glare control can be obtained if strips are printed on both sides of a double-glazed cavity.

The glazing system can also produce energy if the strips are made with photovoltaic thin film.

Prismatic glass

Prismatic glass consists of rolled glass, 3.2–6.4 mm thick, with one face shaped in parallel prisms that refract the transmitted light, thereby changing the direction of the light rays. Under certain conditions of incision and sun height, the rays can also be reflected, thus making prismatic glass a shading element that still allows a certain amount of diffuse radiation to penetrate into the building.

Prismatic glass can be used as a fixed solar control device, especially when placed in a glass cavity, or as adjustable louvers that also allow views out. It does not allow views out because of its light diffracting properties.

Light tubes

Light tubes, or light pipes, are used for transporting or distributing natural or artificial light. A tube lined with highly reflective material leads the light rays through a building, starting from an entrance-point located on its roof or one of its outer walls. The entrance point usually comprises a dome, which has the function of collecting and reflecting as much sunlight as possible into the tube. Many units also have a directional "collectors", "reflectors" or even Fresnel lens devices that assist in collecting additional directional light down the tube.

Light transmission efficiency is greatest if the tube is short and straight. In longer, angled, or flexible tubes, part of the light intensity is lost. To minimize losses, a high reflectivity of the tube lining is crucial; manufacturers claim reflectivity of their materials, in the visible range, of up to 99 %. At the end point (the point of use), a diffuser spreads the light into the room. Devices using optical fibres to transport daylight are also under development, with interesting perspectives in refurbishment thanks to their very small diameter.

In view of the small dimension of the fibres, an efficient daylighting set-up requires a parabolic collector to track the sun and concentrate its light. Light pipes can be also used in new construction and refurbishment, to transport daylight through thick roof structures and attics.

Heliostats
A heliostat tracks the movement of the sun. It is typically used to orient a mirror, throughout the day, to redirect sunlight along a fixed axis towards a stationary target or receiver, which can be an interior building space, a light pipe or an atrium.

In the case of daylight applications, a heliostat can be used to improve the amount of natural light in very deep, very high or scarcely illuminated building areas. In the case of building applications, a heliostat is typically composed of a movable mirror that follows the movement of the sun, and a fixed mirror oriented towards the interior of the building.

Example: Heelis office building, Swindon, UK Heelis, the new headquarters of the National Trust in Swindon, materializes the relationship between natural light, solar shading, ventilation and thermal storage. Heelis received the "excellent" Building Research Establishment Environmental Assessment Method (Breeam) scoring. The building shows the possibility to integrate energetic strategies with low cost technological solutions.
Some of the innovative solutions include:

- natural light for solar passive gain (Victorian shed);
- indirect natural light integrated with an artificial fluorescent system controlled by sensors that
- keep the minimum luminance;
- photovoltaic panels;
- natural ventilation by stack effect (light metal snouts);
- lightweight steel structure (Fig. 3.8).

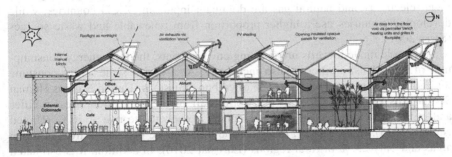

Fig. 3.8 Heelis: A section of the building showing the indirect lighting system and the natural ventilation strategies (Feilden Clegg Bradley Studios).
Design (Arup Associates). Feilden Clegg Bradley Studios; 2005

3.2 Energy Generation from Renewable Sources

Overview

This section considers renewable generation in urban areas both at building and district levels, highlighting barriers and areas of innovation for renewable technologies.

The direction of this report has been shaped by stakeholder input in the Smart-ECO project, including direct survey of industry experts and findings from interactive workshops that took place during the development of the research.

On the topic of renewable energy generation, the key opinions expressed by stakeholders were:

- no European-wide renewable solution, but systems need to be designed to suit local conditions focussing on multiple sources at building and district level;
- the potential of existing technologies is not being fully utilised. Both improving existing and developing new technologies is important, but the focus should be on implementation and integration of existing technologies;
- the most important issues to help improve innovation are applied R&D and exchange of know how. This is considered a higher priority than policies and market measures;
- existing buildings pose the greatest challenge and priority should be given to residential buildings and offices;
- renewable energy technologies are the primary aspect towards the creation of Smart-ECO solutions (in particular solar thermal, earth energy and co-generation).

Using the stakeholder views as a starting point, this section aims to identify areas of innovation with the greatest potential to impact Smart-ECO buildings in the period 2010–2030.

Understanding the problems

Sources of energy used in buildings vary significantly based on economic development: electricity and natural gas are dominant inputs in OECD countries, while developing countries use a higher proportion from renewable and waste sources (Stern 2007).

As people move towards urban living environments, the total energy consumption of high density cities is also increasing. At a global scale, cities consume 80 % of the energy and release 75 % of carbon emissions (United Nations Human Settlements Program 2012). In the world, half of the population lives in cities and in the next four decades this percentage is expected to rise at 70 % (United Nations 2013). In the EU-27, today around 40 % of the population lives in urban areas and only 25 % in rural regions (Eurostat 2010). In addition, it is estimated that at least 90 % of buildings today will still be standing in 2030 (Tofield and Ingham 2012), hence the main challenge in the next 20 years will be to improve the energy efficiency of the existing building stock in urban areas.

Although there is the potential to increase energy generation at building level, the nearly zero-energy standard required for new buildings at the end of this decade is far more problematic than it may appear. The zero-energy balance (intended here broadly as the ability to generate in a year as much energy is consumed) is possible for new buildings in small to medium residential dwellings in low density areas, as different prototypes now demonstrate; however, for buildings in medium to high density areas, potential renewable generation falls significantly short of consumption (Howard et al. 2012). This is mainly due to a lack of usable area, physical constraints of other buildings, limitations of existing building and the low efficiency of generation. The last of these can and will be improved in the future, but renewable generation is still constrained by intermittent supply and a typical mismatch between energy generation and requirement (both daily and seasonal).

A recent vision for European buildings in 2050 (European Commission 2010b) suggests that, while in low density areas ("park city" and "thinly populated" areas) energy could be supplied entirely by an optimised combination of solar and wind energy, in high density and historical areas—where buildings are protected because of their cultural value—most of the energy will come from elsewhere, with a limited part of the supply being generated and stored locally. In fact, while low and medium density urban areas have the potential to utilize public spaces (squares, gardens, car parks, roads, etc.) in addition to building roofs and envelopes, high density areas are instead characterized by multi-level buildings and limited vacant land and present a great challenge for increasing sustainable energy production at a local level, thus having a higher reliance on grid energy (both for heating and electricity).

Provided that—in the foreseeable future—cities will always rely on external inputs of energy because of their density, it is possible to improve the energy performance of urban areas through a mix of improved efficiency, solar urban design whenever possible (Lobaccaro et al. 2012) and efficient energy generation.

How do we tackle the problem of increasing renewable energy generation in urban environments? And, which technologies have the greatest potential to impact buildings? This section will present some promising solutions for the generation at the building level and for the integration of buildings in wider energy networks.

3.2.1 Areas for Innovation: Buildings

While it is unlikely that all existing and new buildings can be energy autonomous, innovation should focus on improving the efficiency of technologies to minimise the reliance on external energy supply.

The split of renewable energy generated by buildings and local grid will depend on each building location, use, age, mass and fabric/structure. Existing buildings will heavily rely on off-site renewable technologies and imported renewable fuel, while new buildings will have a larger opportunity to maximize the use of on site generation and become exporters to the grid where possible (European Commission 2010b, cit.).

Based on the efficiency (power production per unit of land used), the key renewable source to focus innovation for buildings is the sun, especially solar PV and solar thermal.

While traditional photovoltaic cells are widely used, PV innovation must now focus on improving efficiency to enable a greater proportion of energy to be generated from the relatively small amount of roof space on buildings. Multiple junction solar cells are considered to be one approach of third generation solar cells aiming to substantially increase efficiency, following the widely used single junction devices such as crystalline silicon and thin film cells (Guter et al. 2009). Multiple junction cells consist of multiple thin films, each absorbing part of the electromagnetic radiation passing through the cell, so it captures more of the solar spectrum and thus produces more electricity. By using multiple junctions with several band gaps, different portions of the solar spectrum may be converted by each junction at a greater efficiency.

Building integrated photovoltaic (BIPV) systems are photovoltaic components that are used to replace conventional building materials in parts of the building envelope such as the roof, skylights, or facades. They are increasingly being incorporated into the construction of new buildings as a principal or ancillary source of electrical power, although existing buildings may be retrofitted with BIPV modules as well.

The advantage of integrated photovoltaic systems over more common, non-integrated systems is that the initial cost can be offset by reducing the amount spent on building materials and labour that would normally be used to construct the part of the building that the BIPV modules replace. In addition, since BIPV are an integral part of the design, they generally blend in better and are more aesthetically appealing than other solar options. Transparent PV modules, in particular, allow for integration in glazed facades without limiting the views out (Munari Probst and Roecker 2012). Up to now, BIPV have been based on slightly modified standard PV modules, but research is under way to provide the market with better and more appealing products (Lobaccaro and Wall 2012).

It is also possible to raise the temperature of a fluid including a specific circuit in the envelope of a building, such as the roof. Water delivered by this "active" surface is not hot enough to be used for DHW systems, but it can be used in solar assisted heat pump systems (Andria et al. 2008). In the future it will, in all probability, be possible to use SAHPS instead of regular electric heat pumps with comparable efficiency and a fraction of the primary energy consumption (Stojanović and Akander 2010).

Solar cooling is a promising technology with some available applications using solar thermal energy to drive a cooling process. Two technologies are used: desiccant solar cooling (DSC) and thermal solar cooling. DSC is a thermal cooling process for air conditioning that produces cool air through a combination of evaporation cooling and humidity removal. It works by passing hot and moist air to over common, solid desiccants (like silica gel or zeolite) to draw moisture from the air and make it more comfortable. The desiccant is then regenerated using solar thermal energy to dry it out, in a cost-effective, low-energy-consumption, continuously repeating cycle. A photovoltaic system can power a low-energy air circulation fan, and a motor to slowly rotate a large disk filled with desiccant. Thermal solar cooling, on the other hand, uses solar thermal collectors to provide thermal energy to drive thermally-driven chillers (usually adsorption or absorption chillers). In both cases, solar panels are required to produce hot water at a relatively high temperature (needed to drive the process). Vacuum tubes are often preferred over flat panels because of their higher efficiency. Development in this area is towards small scale plants that are suitable for individual buildings (Grossman 2002).

In addition to solar thermal technologies, an effective solution for space and water heating is ground source heat pumps. GSHPs use the consistent temperature of heat in the ground to provide a year-round source for space heating and in some cases pre-heats water before it goes into a more conventional boiler. Closed-loop or ground-coupled systems provide the most sustainable solution as they do not foul groundwater like open-loop. The bottlenecks to continued increases in GSHP adoption include equipment and component supply and a lack of adequate capability to install GSHP systems—particularly the underground component of the system entailing a series of plastic pipes buried underground (Hughes 2008). New drilling and loop insertion technologies have been developed in the last few years with the aim to facilitate installation and increase deployment (Yang et al. 2010).

Example: CH_2 Melbourne, Australia CH_2 is a visionary building leading the way in sustainable design and facility management. It was commissioned by the City of Melbourne to promote "green building" in oceanic climates (emphasis on cooling). The result is a 10 storey office building for 540 employees which uses 74 % less energy than a standard public office building. The "smartness" and innovation of the design process combined a clever mix of relatively current technologies to produce an extraordinary integrated result.

Some of the innovative solutions include:

- use of passive design solutions including integrated design of passive (orientation, envelope etc.) and active (services) strategies, artificial heat storage devices (PCM), integrated shading + natural light control/transfer and development of hybrid ventilation system;
- adaptable and flexible design including internal fit out (modular/movable walls and finishes, modular services);
- materials: non-toxic, recyclable;
- construction systems: maximize use of prefabricated materials and precast concrete sections;
- water conservation and storage: water-efficient fittings, rainwater harvesting, re-use of grey water, sewer mining;
- building monitoring: automatic adjustment of the control unit, depending on the behaviour of the user and energy consumption monitoring;
- solar thermal: hot water solar collectors;
- PV panels on the roof;
- engage stakeholders and users: web based forum for designers, architects, developers, investors to share information, promote training courses, publications (Figs. 3.9, 3.10).

Example: VELUXlab, Milan, Italy VELUXlab is the first Italian Nearly Zero-Energy Building in a University Campus. It is placed in Bovisa Campus of Politecnico di Milano and it is a new laboratory for research.

It is the renovation of the átika model home for Mediterranean climate, designed for Velux and travelling around Southern Europe between 2007 and 2009. After this period Velux donated it to Politecnico of Milan in order to make out of this a new research centre for the University.

From the outside, VELUXlab looks like a normal building with white plaster finishing, but it is the result of a complex analysis of the climate and of the needs of the building itself.

Fig. 3.9 CH$_2$: The North façade of the building, protecting the glazed windows from the high summer sun (Diana Snape)

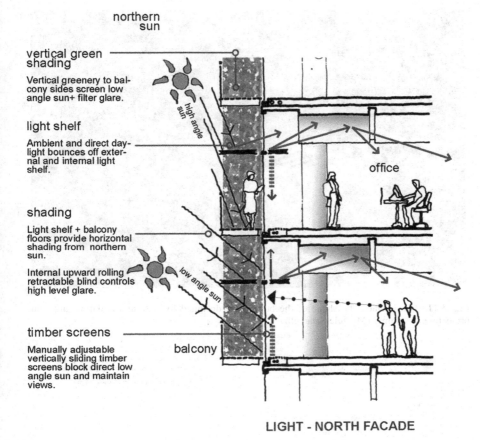

**northern
sun**

**vertical green
shading**

Vertical greenery to bal-
cony sides screen low
angle sun+ filter glare.

light shelf

Ambient and direct day-
light bounces off exter-
nal and internal light
shelf.

shading

Light shelf + balcony
floors provide horizontal
shading from northern
sun.

Internal upward rolling
retractable blind controls
high level glare.

timber screens

Manually adjustable
vertically sliding timber
screens block direct low
angle sun and maintain
views.

high angle sun

low angle sun

office

balcony

LIGHT - NORTH FACADE

Fig. 3.10   CH₂: A scheme of the natural lighting strategies (DesignInc).
Design (Arup Associates). DesignInc; 2006

   The original steel structure was implemented with new materials and insulation levels according to the results of dynamic thermal simulations. Recycled glass
fibre panels were used for cladding and new insulation layers were added. The
indoor comfort was the most important aspect leading the design of the lab: indoor
surface temperatures are relatively high thanks to the insulated and air-tight envelope; ceiling and wall cladding help in cleaning the air thanks to zeolite and gyp-
sum panels; and natural lighting levels are optimised.
   The shape itself is optimized for the warm season, maximising solar gains in
winter and self-shading of windows in summer. The heating energy need of the
building and envelope, without taking into account services and the contribution of
renewable energy sources, is 9.8 kWh/m³a. The building is equipped with a heat-
recovery mechanical ventilation system (90 % efficiency), air-to-water heat pump
(7 kWp) and solar panels for hot water and electricity (2 kWp). This is enough to
balance energy need and production over the year.
The building is now monitored by a team of Politecnico di Milano for comfort levels and energy performances (Figs. 3.11, 3.12).

Fig. 3.11 VELUXlab: The shape of the building is designed for natural ventilation and solar protection in summer (Michele Sauchelli)

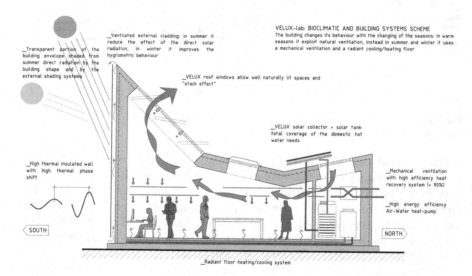

Fig. 3.12 VELUXlab: A scheme of natural ventilation flows (Atelier2—Gallotti e Imperadori Associati and Michele Sauchelli). Design (Arup Associates) of the átika model home: J. A. Cantalejo and A. Ronda, ACXT/IDOM studio; design of VELUXlab: Atelier2—Gallotti e Imperadori Associati; 2012

Example: Ecobox, Madrid, Spain The Fundación Metrópoli in Madrid deals with innovation and sustainability within the design of cities and landscape and it is not a coincidence that their new headquarters, named "Ecobox", is a paradigmatic example of the effects that energy saving issues can have on architecture. The site chosen for the new headquarters is North of the Madrid city centre in the urban expansion area of Alcobendas.

The building's simple parallelepiped shape, which minimizes the surface of the envelope that is exposed to the climate, is internally arranged around an atrium. Taking advantage of the difference in level at the edge of the plot, the designers placed the main entrance at the intermediate level so that when the visitor enters the building he can immediately perceive the lay-out of the different functions.

The efficient passive performance of the building reduces the periods when an active temperature control is required and therefore limits the amount of service requirements to amounts that can be provided almost entirely with renewable energy sources. The integrated approach between architecture and services has allowed for a 60 % saving on the cooling requirements that a traditional office building could require. Some of the innovative solutions include:

- a second external skin made of aluminium blades which automatically controls heat gains depending on the intensity of solar radiation;
- the atrium as the driver of ventilation. Open skylights on top of the building act as chimneys for stack effect ventilation, extracting the exhaust air from the offices. Outside fresh air is pushed by small fans through a concrete labyrinth between the underground slab and the ground, where the air pre-cools or pre-heats;
- the design of the entire building for possible dismantling at the end of its life-cycle;
- vacuum-packed solar collectors on the roof;
- polycrystalline photovoltaic modules on the South façade (Figs. 3.13, 3.14).

Fig. 3.13 Ecobox: The South façade, protecting the glazed portions of the envelope from direct solar radiation and producing energy (Fundación Metrópoli)

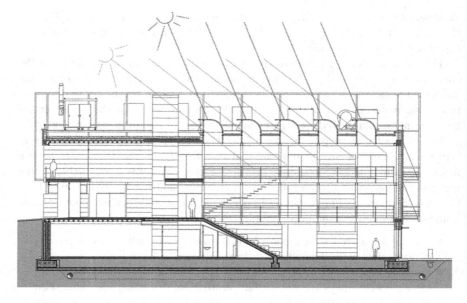

Fig. 3.14 Ecobox: A section of the building, showing the central triple-height atrium (Fundación Metrópoli).
Design (Arup Associates). Ángel De Diego Rica; 2003

Example: New Housing for Seniors, Brescia, Italy The BIRD project (Bioedilizia, Inclusione, Risparmio energetico e Domotica—Green design, Inclusion, Energy saving and Home automation) is divided into a service centre and 52 flats organised in two-storeys rows. Each of the flats has a glazed sunspace, connected to the living rooms, and a loggia or a balcony facing South. The metallic roof abuts the South façade in order to protect it from the high summer sun. Generally speaking, the building includes strategies that prioritise the passive behaviour over the mechanical systems, and produces a large amount of the energy it still needs in the peak seasons.

The building earned an "A+" rating for energy efficiency from the CasaClima Agency, Bolzano, Italy.
Some of the innovative solutions include:

- natural ventilation integrated to mechanical ventilation with heat recovery;
- radiant floor panels for winter heating;
- geothermal heat pumps;
- photovoltaic panels on the roof;
- lightweight steel structure with highly insulated envelope (Figs. 3.15, 3.16).

3.2.2 Areas for Innovation: Integrated Generation

To maximize renewable energy generation, it is important to consider the integration of each building with its surrounding environment and district capability. This

Fig. 3.15 BIRD project: The South façade, with the glazed portions for winter solar collection and the overhanging roof for summer shading (AIACE)

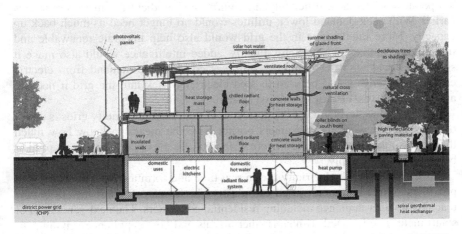

Fig. 3.16 BIRD project: A scheme of the summer strategies to reduce overheating (AIACE). *Design* (Arup Associates). AIACE—Ettore Zambelli with Politecnico di Milano; 2010

relies on significant innovation in the approach to planning building works, monitoring through smart grids and energy storage.

In urban areas, energy savings due to large developments are typically due to the contribution of three main elements: CHP (co-generation), energy efficient and passive design, and use of renewable energy. In urban areas, renewable energy generation at building level has so far contributed only to a small part of reducing CO_2 and that a holistic approach involving district generation is required to satisfy the current emission reduction targets (Day et al. 2009).

To increase uptake, the following issues need to be addressed:

• regulations on the use of on-site and district generation and the interaction between companies in different segments of the energy generation industry;
• holistic town planning strategies to allow sufficient land for sustainable energy generation for new developments;
• all major new developments should establish or connect to a local renewable grid, district heating and district cooling where appropriate;
• tools to assess the capacity of local industry to meet demand and increase the attractiveness of community scale energy industry;
• increasing awareness of the benefits of an integrated approach;
• carbon costs across industry and consumer groups.

3.2.2.1 Smart Grids

Smart grids are being developed to combat the challenge of intermittent energy supply and reduce peak requirements by regulating use (European Commission 2006). At an infrastructure level, sensors fitted to power lines will enable utilities to operate systems more efficiently and reliably and predict transmission problems earlier. With peak demand lower, utilities would no longer need as much back up capacity. More intelligence in the grid would also help integrate renewable and intermittent sources of electricity. Finally, added intelligence would also make it simpler to deal with the imminent and very considerable demand from electric cars. Car batteries could be used to feed electricity back into the grid if needed, and so act as a vast electricity storage system.

At a consumer level, one of the problems of traditional electricity grids is a lack of transparency whereby consumers do not have the means to know how much they are using in real time, and therefore do not have the opportunity to adapt their behaviour to suit supply.

Smart meters track electricity use in real time and form a data connection with providers. While commonly used in industry, the benefit of smart meters is now reaching a wider audience including residential consumers. The next generation of smart meter is integrated with smart thermostats and home appliances giving people more control over how much electricity they are consuming through awareness and automation. As electricity prices rise (and in some cases are charged depending on system load), a home display means consumers can time their usage to be cost effective and also select the source of their electricity.

To facilitate an "energy internet" model, the following need to be addressed:

• sensors and digital relays installed on power lines will enable utilities to operate systems with greater efficiency and reliability;
• large scale roll out of smart meters to establish the two-way data connection between consumer and utility company;

- efficient management of intermittent RES;
- compatible system components to encourage interoperability of devices.

References

D.W. Abrams, *Low Energy Cooling* (Van Nostrand Reinhold, New York, 1986)

H. Akbari, S. Konopacki, Calculating energy-saving potentials of heat-island reduction strategies. Energy Policy **33**(6), 721–756 (2005)

G. Andria, A. Lanzolla, F. Piccininni, G.S. Virk, Design and characterization of solar-assisted heating plant in domestic houses. IEEE Trans. Instrum. Meas. **57**(12), 2711–2719 (2008)

E.P.D. Barrio, Analysis of the green roofs cooling potential in buildings. Energy Build. **27**(2), 179–193 (1998)

G.S. Brager, R.J. de Dear, Thermal adaptation in the built environment: a literature review. Energy Build. **27**(1), 83–96 (1998a)

G.S. Brager, R.J. de Dear, *Developing an adaptive model of thermal comfort and preference*. Center for the Built Environment, UC Berkeley, Berkeley (1998b)

California Energy Commission, *Building energy efficiency standards for residential and non residential buildings*. Sacramento (2008)

T. Day, P. Ogumka, P. Jones, *Monitoring the London Plan Energy Policies—Phase 3—Part 1 Report Final* (London South Bank University, London, 2009)

C. Di Perna, F. Stazi, A. Ursini Casalena, M. D'Orazio, Influence of the internal inertia of the building envelope on summertime comfort in buildings with high internal heat loads. Energy Build. **43**(1), 200–206 (2011)

M.V. Diamanti, M. Ormellese, M. Pedeferri, Characterization of photocatalytic and superhydrophilic properties of mortars containing titanium dioxide. Cem. Concr. Res. **38**(11), 1349–1353 (2008)

Directive 2010/31/UE of the European Parliament and of the Council on the energy performance of buildings (recast)

Directive 2012/27/EU of the European Parliament and of the Council of 25 October 2012 on energy efficiency, amending Directives 2009/125/EC and 2010/30/EU and repealing Directives 2004/8/EC and 2006/32/EC

B. Dunster, C. Simmons, B. Gilbert, *The ZEDbook* (Taylor & Francis, Abingdon, 2008)

Energy Saving Trust, *Fabric first—focus on fabric and services improvements to increase energy performance in new homes*. London (2010)

European Commission, *European technology platform smart grids: Vision and strategy for Europe's electricity networks of the future*. Office for Official Publications of the European Communities, Luxembourg (2006)

European Commission, *Energy-efficient buildings PPP multi-annual roadmap and longer term strategy*. Publications Office of the European Union, Luxembourg (2010a)

European Commission, *Europe 2020—a strategy for smart, sustainable and inclusive growth*, COM(2010) 2020 final. Brussels (2010b)

European Commission, *Energy roadmap 2050—communication from the commission to the European Parliament, the Council, the European economic and social committee and the committee of the regions*, COM(2011) 885 final. Brussels (2011)

European Union, *Energy 2020—a strategy for competitive, sustainable and secure energy*. Publications Office of the European Union, Luxembourg (2011)

Eurostat, *Eurostat regional yearbook 2010*. Publications Office of the European Union, Luxembourg (2010)

Eurostat, *Smarter, greener, more inclusive? Indicators to support the Europe 2020 strategy*. Publications Office of the European Union, Luxembourg (2013)

S. Ferrari, G. Masera, D. Dell'Oro, Improving comfort and energy efficiency in a nursery school design process, in *Proceedings of PLEA 2006—23rd International Conference on Passive and Low Energy Architecture*, Geneva, 6–8 Sept 2006, pp. II425–II430 (2006)

M.P. Fiori, R. Paolini, T. Poli, Monitoring of eight green roofs in Milano. Hygrothermal performance and microclimate mitigation potential, in *39th World Congress on Housing Science. Changing Needs, Adaptive Buildings, Smart Cities*, Milan, 17–20 Sept 2013, pp. 1365–1372 (2013)

F. Frontini, T. Kuhn, Una nuova facciata verticale selettiva per il controllo solare, *Sostenibilità e innovazione in edilizia* ed. by V. Masera, M. Ruta (Aracne editrice, Rome, 2009), pp. 221–233

A.D. Galasiu, J.A. Veitch, Occupant preferences and satisfaction with the luminous environment and control systems in daylit offices: a literature review. Energy Build. **38**(7), 728–742 (2006)

R. Gonzalo, K.J. Haberman, *Energy-Efficient Architecture. Basics for Planning and Construction* (Birkhäuser, Basel, 2006)

G. Grossman, Solar-powered systems for cooling, dehumidification and air-conditioning. Sol. Energy **72**(1), 53–62 (2002)

W. Guter, J. Schone, S.P. Philipps, M. Steiner, G. Siefer, A. Wekkeli, E. Welser, E. Oliva, A.W. Bett, F. Dimroth, Current-matched triple-junction solar cell reaching 41.1 % conversion efficiency under concentrated sunlight. Appl. Phys. Lett. **94**(22), 223504-1–223504-3 (2009)

M. Hegger, M. Fuchs, T. Stark, M. Zeumer, *Energy Manual* (Birkhäuser—Edition Detail, Basel, 2008)

T. Herzog, Solar architecture, in *Energy Manual*, ed. by M. Hegger, M. Fuchs, T. Stark, M. Zeumer (Birkhäuser—Edition Detail, Basel, 2008)

B. Howard, L. Parshall, J. Thompson, S. Hammer, J. Dickinson, V. Modi, Spatial distribution of urban building energy consumption by end use. Energy Build. **45**, 141–151 (2012)

P.J. Hughes, *Geothermal (Ground-Source) Heat Pumps: Market Status, Barriers to Adoption, and Actions to Overcome Barriers*. US Department of Energy Publications (2008)

M. Imperadori, G. Masera, G. Iannaccone, D. Dell'Oro, Improving energy efficiency through artificial inertia: The use of phase change materials in light, internal components, in *Proceedings of PLEA 2006—23rd International Conference on Passive and Low Energy Architecture*, Geneva, 6–8 Sept 2006, pp. I547–I552 (2006)

International Energy Agency, *Technology roadmap—energy efficient buildings: Heating and cooling equipment* (2011)

G. Lobaccaro, M. Wall, *Product developments and dissemination activities*. International Energy Agency—Solar Heating and Cooling Programme, Task 41—Solar Energy & Architecture (2012)

G. Lobaccaro, F. Fiorito, G. Masera, T. Poli, District geometry simulation: a study for the optimization of solar façades in urban canopy layers. Energy Proc. **30**, 1163–1172 (2012)

M.C. Munari Probst, C. Roecker, *Solar energy systems in architecture: Integration criteria and guidelines*. International Energy Agency—Solar Heating and Cooling Programme, Task 41—Solar Energy & Architecture (2012)

F. Rubel, M. Kottek, Observed and projected climate shifts 1901–2100 depicted by world maps of the Köppen-Geiger climate classification. Meteorol. Z. **19**(2), 135–141 (2010)

I. Sartori, A.G. Hestnes, Energy use in the life cycle of conventional and low-energy buildings: a review article. Energy Build. **39**(3), 249–257 (2007)

N. Stern, *The Economics of Climate Change: The Stern Review* (Cambridge University Press, Cambridge, 2007)

B. Stojanović, J. Akander, Build-up and long-term performance test of a full-scale solar-assisted heat pump system for residential heating in Nordic climatic conditions. Appl. Therm. Eng. **30**(2–3), 188–195 (2010)

B. Tofield, M. Ingham, *Refurbishing Europe—An EU strategy for energy efficiency and climate action led by building refurbishment*. UEA Low Carbon Innovation Centre and Build with CaRe (2012)

T. Turney, Future materials and performance, in *Technology, Design and Process Innovation in the Built Environment*, ed. by P. Newton, K. Hampson, R. Drogemuller (Taylor & Francis, Abingdon, 2009)

United Nations Human Settlements Programme, *State of the World's Cities 2012/13* (UN-HABITAT, Nairobi, 2012)

United Nations, Department of Economic and Social Affairs, Population Division, *World population prospects: The 2012 revision, highlights and advance tables*. Working Paper No. ESA/P/WP.228 (2013)

K. Voss, E. Musall, *Net Zero Energy Buildings—International Projects of Carbon Neutrality in Buildings* (Institut für Internationale Architektur-Dokumentation, Munich, 2011)

M. Vraa Nielsen, S. Svendsen, L. Bjerregaard Jensen, Quantifying the potential of automated dynamic solar shading in office buildings through integrated simulations of energy and daylight. Sol. Energy **85**(5), 757–768 (2011)

H. Yang, P. Cui, Z. Fang, Vertical-borehole ground-coupled heat pumps: a review of models and systems. Appl. Energy **87**(1), 16–27 (2010)

Chapter 4
Meeting Future Requirements

Abstract The chapter focuses on the identification of changes that will affect the building sector in the next future. The following areas of concern were identified as the most critical ones: energy and environment, changing requirements and depletion of resources. The following part identifies three topics that are expected to impact on buildings and the way we conceive, build and manage them. Of course, it is extremely difficult to predict what will be the future in the next 20 years or so; nonetheless, it is possible to point out a few aspects that designers and decision-makers should consider to deliver buildings that can face future challenges.

Keywords Climate change · Adaptive buildings · Life cycle · Adaptable and flexible design · Smart grid · Design for deconstruction

The design of buildings has always been a complex matter, not least because life cycles are extremely long and it is difficult to predict what will happen in decades. Today, this challenge is made even more compelling because climate change is accelerating and we may find ourselves in 2030 with a very different world from the one we have known in the last century.

Buildings constructed today will very probably be still in use when fossil fuels will be no longer available and should be ready to be retrofitted for other forms of energy supply. Buildings should invest more on the "passive" behaviour rather than on "eco-bling" technologies applied for greenwashing purposes (Liddell 2013).

Flexibility and adaptability of buildings (not limited to internal fit-out, but including the external envelope) seems to be a crucial point if our buildings are to face uncertain conditions of climate and use. On the other hand, escalating costs for materials (due to scarcity of resources) and maintenance (due to increased stress on the building as a consequence of climate change) may push in the direction of self-healing structures and materials, involving nanotechnologies and surface science (Turney 2009).

Buildings will very likely become (the trend is already underway) more and more akin to organisms, with feedback loops controlling its behaviour, managing energy

This chapter is written by Giuliana Iannaccone and Gabriele Masera

© The Author(s) 2014
G. Iannaccone et al., *Smart-ECO Buildings Towards 2020/2030*,
PoliMI SpringerBriefs, DOI 10.1007/978-3-319-00269-9_4

flows efficiently, changing the functionality of the envelope according to external conditions and comfort requirements from the users (Turney 2009). Whether this will happen through innovative materials, "smart" control systems or a combination of both, the psychological effect of these far-fetching ideas on the occupants should not be underestimated, and a degree of control (and responsibility) should be left to the users, as pointed out in Sect. 3.1.1.6.

4.1 Energy and Environment

4.1.1 Mitigation and Adaptation for Climate Change

Awareness of climate change and the contribution of buildings to energy use, resource consumptions and CO_2 emissions is widespread (Santamouris 2001).

Some effects of the warming planet are already being felt, and further consequences are on their way. These changes will vary from region to region (Sanders and Phillipson 2003), but general trends include changing precipitation patterns and heavier downpours, even in areas where overall precipitation will decline; longer, hotter, and more frequent heat waves; rising sea levels due to melting glaciers and land-based ice sheets; loss of both sea ice and protective snowpack in coastal areas; stressed water sources due to drought and decreased alpine snowfall; and "positive feedback loops"—consequences of warming that cause further warming, such as melting sea ice decreasing the capacity of the northern oceans to reflect solar radiation back out of the atmosphere.

Climate change brings new challenges, which impact on the natural and built environments and aggravates existing environmental, social and economic problems. These changes will affect different aspects of spatial planning and the built environment, including external building fabric, structural integrity, internal environments, service infrastructure, open spaces, human comfort and the way people use indoor and outdoor space. Coupled with the challenges of rapid urbanization, climate change impacts may undermine country efforts to achieve the goals of sustainable development (Luebkeman 2007).

Climate change policy has developed around two themes: *mitigation* and *adaptation*. Mitigation is tackling the causes of climate change through reduction of greenhouse gas emissions; adaptation is adjusting to the physical impacts of climate change, by reducing vulnerability and finding opportunity. Mitigation and adaptation should not be viewed as alternatives. Adaptation will be needed to deal with the unavoidable impacts of climate change even with mitigation. In the longer term, adaptation is likely to be insufficient to manage the most serious impacts of climate change should mitigation efforts fail (Stern 2006; ARUP 2008). Expected changes to central Europe, for example, include increase in average temperature, more heat waves and droughts in summer, milder, wetter winters, more winter storms and more frequent extreme conditions throughout the year (Werner and Chmella-Emrich 2009).

The building sector has a considerable potential for positive change both in mitigation and adaptation strategies, to become more efficient in terms of resource use and environmental impact. Considering the effects of climate change, building practices will have to change to ensure buildings continue to fulfill their functions throughout their life cycle.

Most of the strategies for adapting buildings to the effects of climate change are described in this book (see Chap. 3). The adaptive measures listed below give us something we can think about and act upon today. The good news is that many of these measures also help to mitigate climate change—and quite a few reduce building operating costs or improve durability, benefiting building owners as well as the future of the planet.

The challenge to achieving sustainable buildings and reduced climate change impact is usually not a lack of access to technical solutions but a lack of uptake by building sector stakeholders. This challenge must be tackled through policy, finance and education.

4.1.2 Integration of Buildings in the Energy Networks

Consideration of energy efficiency in buildings needs to be embedded in considerations of energy efficiency on an urban scale (or rural or suburban), including influence on traffic patterns as well as on energy infrastructure (UNEP 2007). Trends emerging in the power system suggest that the highly centralized paradigm that has dominated power systems for the last century may eventually be replaced, or at least diluted, by an alternative. Researchers worldwide are recognizing the promise of micro-grids to improve energy efficiency by moving thermal generation close to possible uses. This would permit waste heat recovery and use, and better integration of small-scale dispersed renewables into the energy supply infrastructure.

De-centralised supply plants, in fact, allow for more economical and efficient coupling of systems with alternative energy sources (wind and sun, with "distributed availability") and for the use of heat sources within the development area (groundwater, air, waste heat) (Gonzalo and Haberman 2006). This concept improves the safety of energy supply, which, at our present state of knowledge, means using the energy available at the site and avoiding import as far as possible (exceptions were highlighted in Sect. 3.2) (Hegger et al. 2008).

Buildings are very likely to become "virtual power stations" producing surplus energy and feeding it into the smart grids that were presented in Sect. 3.2. The challenge for the future is the integration of centralised and de-centralised sources of energy, balancing demand and supply (which is intermittent due to the nature of renewable sources). Energy will have to be stored with innovative means, such as the fuel cells (as discussed earlier) or electric cars plugged to the grid. Plug-in hybrid fuel cells for cars may contribute electricity and heat when the car is parked and connected to a building (and parked it is most of the time). Fundamental research shows possibilities to directly utilize the energy rich sugar alcohols (sorbitol, mannitol) derived via a catalytic process from cellulose, hemi-cellulose of rapidly growing leaf trees (e.g. aspen, poplar, hybrid-poplar, etc.).

The potential of this new approach to integration of buildings and transport into a distributed energy network is seen by some as a "third industrial revolution" (Rifkin 2014).

This power supply evolution poses new challenges to the way buildings are designed, built, and operated. Traditional building energy supply systems will become much more complex in at least three ways.

First, architects and building engineers cannot assume that as now gas will arrive at the gas meter, electricity at its meter, and within the structure, the two systems are virtually independent of one another. Rather, energy conversion, heat recovery and use, and renewable harvesting may all be taking place simultaneously at various locations within the building energy system. Second, the structure of energy flows in the building must accommodate multiple energy processes in a manner that permits high overall efficiency. In other words, the building must be designed around its energy flows and energy equipment to ensure efficiency. And third, multiple qualities of electricity may be supplied to various building functions, and there placement and supply must be considered (Marnay and Firestone 2007).

4.1.3 Building Strategies to Reduce Depletion of Resources

Increasing scarcity and the consumption of fertile land and natural resources are a significant global problem.

The use of building materials should be reduced considerably as a means of resource efficiency. In order that materials remain available permanently, open materials chain, especially those for non renewable raw materials, must be closed wherever possible. Actually, the strategies to use materials efficiently and to integrate building materials in closed cycles are being applied sporadically. But the recent Construction Product Regulation (05/2011) has introduced a basic requirement on sustainable use of natural resources that will demand proof of the environmental impact of building materials in accordance with the life cycle assessments.

The extension of materials life-cycle could be obtained by promoting the extension of life expectancy in buildings, i.e. by means of conversion/transformation of existing buildings instead of new construction.

As explained in a previous chapter, the largest part of the embodied energy is in the load-bearing structure. The refurbishment of existing building enable further use of load-bearing structure and a great potential for resource and energy savings.

In a flexible or adaptive building, growing usage expectations and correspondingly better internal fitting-out mean that long-lasting components are not fully exploited. In such cases the design should take into account renewal processes and, if feasible, secondary uses for components and materials.

Considering the life cycle of buildings means the design of building materials, components, information systems, and management practices to create buildings that facilitate and anticipate future changes to and eventual adaptation or dismantling for recovery of all systems, components, and materials.

Buildings would be designed for deconstruction and built using materials recovered from other buildings that themselves had been designed for deconstruction. Even if it is not possible to achieve a completely closed-loop building the life cycle (i.e., to eliminate the need for any new materials or systems), adhering to lifecycle construction principles whenever possible can provide meaningful benefits by reducing the energy and resource consumption required to produce the necessary building materials and systems and by reducing solid waste.

In the near term, three specific lifecycle construction practices will offer the greatest potential (see Sect. 2.2):

- Deconstruction of older buildings that were not designed for deconstruction, with reuse of salvaged materials in other building projects whenever possible;
- Design for Deconstruction (DfD) and Materials Reuse—Construction of new buildings using DfD principles and, where possible, incorporating salvaged building materials (Kibert 2003).
- Retrofitting and new construction of buildings to include green building elements such as sustainable site planning, energy efficiency, safeguarding water and water efficiency, conservation of materials and resources, and improved indoor environmental quality.

4.2 Changing Requirements

4.2.1 Meeting Future Performance Levels

Sustainable design requires that service life considerations be integrated in the design process. The preferable outcome is for Smart-ECO buildings to have a long service life, thereby minimizing the drain on resources and attendant environmental effects. One factor in achieving that outcome is to design with flexibility and adaptability as key criteria. It is equally important to design for deconstruction, maximizing reuse and recycling potentials, when the building is not likely to stand for a long time for one reason or another.

Post-occupancy evaluation (POE) and facility performance evaluation (FPE) are continuous processes of systematically evaluating the performance and/ or effectiveness of one or more aspects of buildings in relation to issues such as accessibility, aesthetics, cost-effectiveness, functionality, productivity, safety and security, and sustainability.

There are several potential short, medium, and long-term benefits of FPEs:

- short-term benefits are related to immediate design decisions, and facility maintenance and management issues;
- medium-term (3- to 5-years period) benefits are related to generating useful information for future projects, and
- long-term (10- to 25-years period) benefits are related to improving the long-term performance of buildings and to justifying major expenditures.

4.2.2 Adaptable and Flexible Design for Future Needs

Changes characterizing our society include an ageing population, urban migration, our lifestyle and work (UNDP 2013). These often make traditional building approaches obsolete.

The existing building stock cannot totally satisfy the changed needs and new projects ask for careful valuations and new operating tools. To face the change, a feasible solution is to introduce in an architectural project the requisite of flexibility: i.e. to realize adaptive buildings, or buildings that can modify their characteristics according to changing boundary conditions.

If some classes of buildings can be expected to have a long useful life, other classes may have moderate rates of adaptation because of social and technical change (e.g. housing, schools, and industrial buildings) and others (office, commercial and health care facilities) are experiencing a fast change. Adjustments to existing buildings in order to accommodate new functions or technical systems can be disruptive and wasteful. Lessons from these experiences of adapting existing stock are beginning to find use in the design of new, adaptable buildings. This implies the development of innovative design and construction methods, new products, new regulatory and financing schemes and new performance assessment tools.

A design based on the adaptability and flexibility concept makes it possible to continue using the building even if needs have changed: this is the *loose fit, long life* concept, that aims at the maximum reuse of the structural components of the building—structural frames and floor slab embody, on average, 50 % of the grey energy of a building (Hegger et al. 2008).

Compared to the high profile nature of low carbon design, adaptability (to mean adaptable design) is still a minor agenda, being confined mainly to academic study and application in limited case studies (Grinnell et al. 2011).

For buildings to be adaptable they should be able to accommodate substantial changes. As most buildings are designed for a considerable lifespan it is inevitable that changes will be required. This is especially valid in relation to building materials and services which are developing all the time. A building that is adaptable will be utilized more efficiently and may stay in service longer as it can respond to change at a lower cost. A longer service life may in turn translate to a better environmental performance over its lifecycle.

Innovative solutions are required to ensure that new buildings are flexible and adaptable to future change.

The following points illustrate the strategies for adaptive/flexible buildings.

- Structural efficiency of buildings can be increased by using frames, braced in a proper way, usually in steel or wood. The embodied energy of such solutions is lower than massive brick/concrete or concrete frames and this also allows to save money and materials for foundations. Steel or wood frame structures can be reused/recycled in the future.
- Fixed parts like staircases, balconies, ramps, can be located outside the inhabited box. If this is made in a steel/wood frame system, it will allow a great flexibility in the interior transformation which could be obtained during the service life cycle

of the building. Using internal lightweight walls, flexibility and transformation can be easily obtained without using too much energy for the necessary works.

• Adaptable internal fit out (modular/movable walls and finishes, modular services). Walls with fittings and mobile walls allow for simple and 'soft' changes in the internal fit-out of spaces according to changes in uses ad habits. In the long term, modular services, with smart plug-ins, could help the internal flexibility.

Flexible façade technology that allows for change in use (e.g. offices vs. housing). Façade elements are designed and conceived with the same level of flexibility as the interior walls. Outdoor spaces could be integrated in the interior spaces (e.g. balcony could change into loggia). A right evaluation of changes in natural lighting, fire safety, noise reduction, solar gains, thermal insulation is needed in order to guarantee always high levels of comfort.

References

ARUP, *Drivers of Change* (Prestel, Munich, 2008)

R. Gonzalo, K.J. Haberman, *Energy-Efficient Architecture. Basics for Planning and Construction* (Birkhäuser, Basel, 2006)

R.C. Grinnell, S.A. Austin, A.J. Dainty, Reconciling low carbon agendas through adaptable buildings, in *Proceedings of the 27th Annual ARCOM Conference*, ed. by C. Egbu, E.C.W. Lou, Bristol, UK, 5–7 Sept 2011

M. Hegger, M. Fuchs, T. Stark, M. Zeumer, *Energy Manual* (Birkhäuser, Basel, 2008)

C.J. Kibert, Deconstruction: the start of a sustainable materials strategy for the built environment. UNEP Ind. Environ. **26**(3), 84–88 (2003)

H. Liddell, *Eco-minimalism. The Antidote to Eco-bling* (RIBA Publications, London, 2013)

C. Luebkeman, Global change, in *Energy Manual*, ed. by M. Hegger, M. Fuchs, T. Stark, M. Zeumer (Birkhäuser, Basel, 2007), pp. 10–13

C. Marnay, R.M. Firestone, Microgrids: an emerging paradigm for meeting building electricity and heat requirements efficiently and with appropriate energy quality. In: European Council for an Energy Efficient Economy 2007 Summer Study, 4–9 June 2007. La Colle sur Loup, France (2007). http://eetd.lbl.gov/sites/all/files/publications/conference-paper-lbnl-62572.pdf. Accessed 23 Dec 2013

J. Rifkin, The Third Industrial Revolution (2014). http://www.thethirdindustrialrevolution.com. Accessed 2 May 2014

C.H. Sanders, M.C. Phillipson, UK adaptation strategy and technical measures: the impacts of climate change on buildings. Building Research & Information **31**(3–4), 210–221 (2003)

M. Santamouris (ed.), *Energy and Climate in the Urban Built Environment* (James & James, London, 2001)

N. Stern, What is the economics of climate change? World Econ. **7**(2), 1–10 (2006)

T. Turney, Future materials and performance, in *Technology, Design and Process Innovation in the Built Environment*, ed. by P. Newton, K. Hampson, R. Drogemuller (Taylor and Francis, Abingdon, 2009), pp. 31–53

UNDP, Human Development Report. The Rise of the South: Human Progress in a Diverse World (2013). http://hdr.undp.org/sites/default/files/reports/14/hdr2013_en_complete.pdf. Accessed 23 Dec 2013

UNEP, Buildings and Climate Change. Status, Challenges and Opportunities, UNEP, Paris (2007). http://www.unep.fr/shared/publications/pdf/DTIx0916xPA-BuildingsClimate.pdf. Accessed 23 Dec 2003

P. Werner, E. Chmella-Emrich, Building for climate change—Global warming and its consequences. Detail Green **02**, 14–17 (2009)

Errata to: Smart-ECO Buildings Towards 2020/2030

Giuliana Iannaccone, Marco Imperadori and Gabriele Masera

Errata to: G. Iannaccone et al., *Smart-ECO Buildings Towards 2020/2030*, SpringerBriefs in Applied Sciences and Technology, DOI 10.1007/978-3-319-00269-9

Below listed corrections are need to be incorporated in published volume:

Page No	Published content	Replace with
Page 13: Author name in the article note	"This chapter is written by Gabriele Masera"	"This chapter is written by Giuliana Iannaccone"
Page 37: Author name in the article note	"This chapter is written by Giuliana Iannaccone"	"This chapter is written by Gabriele Masera"

The online version of the original book can be found under
DOI 10.1007/978-3-319-00269-9

Giuliana Iannaccone (✉)
Marco Imperadori
Gabriele Masera
A.B.C. Department, Politecnico di Milano, Milan, Italy

© The Author(s) 2014
G. Iannaccone et al., *Smart-ECO Buildings Towards 2020/2030*,
PoliMI SpringerBriefs, DOI 10.1007/978-3-319-00269-9_5

Index

© The Author(s) 2014
G. Iannaccone et al., *Smart-ECO Buildings Towards 2020/2030*,
PoliMI SpringerBriefs, DOI 10.1007/978-3-319-00269-9

Printed in the United States
By Bookmasters